D0882550

RUDOLF DIESEL: *Pioneer of the Age of Power*

Rudolf Diesel

Pioneer of the Age of Power

BY W. ROBERT NITSKE
AND CHARLES MORROW WILSON

*UNIVERSITY
OF OKLAHOMA PRESS
NORMAN*

Frontispiece: First Diesel engine used
commercially in the United States

Rudolf Diesel has been printed on paper bearing the watermark of
the University of Oklahoma Press and having a life expectancy of
more than three hundred years.

LIBRARY OF CONGRESS CATALOG CARD NUMBER: 65–10110

Copyright 1965 by the University of Oklahoma Press, Publishing Division of the
University. Composed and printed at Norman, Oklahoma, U.S.A., by the
University of Oklahoma Press. First edition.

Acknowledgments

THE AUTHORS are indebted to the following organizations and companies, institutions, and individuals who supplied material which was used in this book:

Aktieselskabet Burmeister & Wain's *Maskin- og Skibsbyggeri,* Copenhagen, Denmark.

Anciens Établissements Sautter-Harlé, Paris, France.

Association of American Railroads, Washington, D.C.

Associazione Nazionale Fra Industrie Automobilistiche E Affini, Turin, Italy.

Baldwin-Lima-Hamilton Corporation, Hamilton, Ohio.

William C. Balmanno, Sunnyvale, California.

Belge du Commerce Extérieur, Brussels, Belgium.

British Internal-Combustion Engine Manufacturers' Association, London, England.

Buda Division, Allis-Chalmers Manufacturing Company, Harvey, Illinois.

Caterpillar Tractor Company, Peoria, Illinois.

Cleveland Diesel Engine Division, General Motors Corporation, Cleveland, Ohio.

Le Conseiller Commercial de France, San Francisco, California.

Cooper-Bessemer Corporation, Mount Vernon, Ohio.

Daimler-Benz, A.G., Stuttgart-Untertürkheim, Germany.

Diesel Progress, Los Angeles, California.

Fachgemeinschaft Kraftmaschinen im VDMA, Frankfurt, Germany.

Fairbanks-Morse and Company, Chicago, Illinois.

Frankfurter *Allgemeine Zeitung,* Frankfurt, Germany.

A/S Frichs, Aarhus, Denmark.

Klöckner-Humboldt-Deutz, A.G., Cologne, Germany.

Fried. Krupp *Dieselmotoren G.m.b.H.,* Essen, Germany.

Library of Congress, Washington, D.C.

Machinefabriek Werkspoor, N.V., Amsterdam, Netherlands.

Maschinenfabrik-Augsburg-Nürnberg, Augsburg, Germany.

Mirrlees, Bickerton & Day Ltd., Hazel Grove, Cheshire, England.

Nordberg Manufacturing Company, Milwaukee, Wisconsin.

Nydqvist and Holm *Aktiebolag,* Trollhätten, Sweden.

Max J. B. Rauck, Munich, Germany.

Ruston and Hornsby Ltd., Lincoln, England.

Santa Barbara (Calif.) Public Library.

Dr. Friedrich Schildberger, Stuttgart-Vaihingen, Germany.

Ellis M. Sims, University of Oklahoma, Norman, Oklahoma.

Société Anonyme Adolph Saurer, Arbon, Switzerland.

Georg Strössner, *Oberingenieur,* Augsburg, Germany.

Süddeutsche Zeitung, Munich, Germany.

Sulzer Brothers Ltd., Winterthur, Switzerland.

Swiss Association of Machinery Manufacturers, Zurich,
  Switzerland.

Franco Tosi, *Società per Azioni,* Legnano, Italy.

University of California Library, Berkeley, California.

Fürst Albrecht von Urach-Württemberg, Stuttgart, Germany.

Waukesha Motor Company, Waukesha, Wisconsin.

Otto Wiedner, St. Nikolaus, Saarland, Germany.

Worthington Corporation, Harrison, New Jersey.

Contents

Acknowledgments   v

Overture   3

I. The Orphans of Paris   5

II. Industrial Revolution   26

III. Song of the Homeland   38

IV. Young Man Makes Good   53

V. Her Name Was Martha   59

VI. The Serious Inventor   72

VII. *"Wunderbar"*   83

VIII. Back to Work   92

IX. An Engine Is an Engine Is an Engine   104

X. Laurel for the Victor's Brow, Ashes for His Cheeks   113

XI. "Build a Better——"   124

XII. Theory *Versus* Workability   136

XIII. The Uncertainly Rich Man   145

XIV. A House, a Philosophy, and an International Success   157

XV. The Black Mistress Goes International   172

XVI. The Inventor as Investor   191

XVII. Herr and Frau Diesel Go to St. Louis   203

XVIII. Side Trip: A Remembrance
by Charles Morrow Wilson   216

XIX. Farewell to America   226

xx. *"En attendant je t'aime"* 235
xxi. Diesel Engines Throughout the World 250
xxii. The Diesel Engine Aloft 278
Finale 284
Chapter Bibliographies 285
A Diesel Bibliography 293
Index 303

Illustrations

🛆
🛆
🛆 *Between pages 118–19*

Frontispiece: First Diesel engine used commercially in the
    United States
Rudolf Diesel
Martha Diesel
Diorama of the Diesel workshop
Munich exhibition display of the Diesel engines
First Diesel engine in Great Britain
A Diesel engine eliminating the crosshead
First Diesel engine used commercially in the United States
A two-cylinder Diesel marine engine
The engine built for the submarine *Deutschland*
First large stationary Diesel installation
Three-cylinder Type A Diesel engine

🛆 *Between pages 230–31*

M.A.N.'s first Diesel automotive engine
Diesel-powered earthmover
Caterpillar Diesel D8 tractor
Diesel-electric tree crusher
Diesel engine at irrigation pumping station
Power-placement diagram for a modern Diesel locomotive
Largest Diesel engine built in the Western Hemisphere
Rudolf Diesel

RUDOLF DIESEL: *Pioneer of the Age of Power*

Overture

In the laboratory, which smelled strongly of oil and hot machines and burnt gases, the unsteady rhythm of the working engine sounded above the noisy clatter of the busy factory which surrounded it.

Then, for the moment, the constant slapping of the wide leather belt on the flywheel appeared to have been interrupted. The big wheel gave an abrupt jerk and pulled the overhead transmission instead of the leather belt. The workman atop the wooden platform surrounding the tall engine lifted his cap in a silent tribute to the occasion which he knew was momentous. The smock-clad young inventor standing below recognized the salute. He reached up and shook his helper's hand. The great work seemed crowned with success. There now appeared to be no doubt at all that the name Rudolf Diesel would be assured a place in history. The Diesel engine, destined to be a prime mover for what is called civilization, was working and generating power.

The Orphans of Paris

THE YOUNGISH MAN, who did not look or act young, and his exceptionally handsome nine-year-old son were among thousands of Parisians out for the traditional week-end stroll in the country.

The year was 1867. Paris was neither young nor especially gay. But her countryside strollers were, as ever, irrepressible. In the main, and as usual, they were disposed to avoid the obvious. On this particular Sunday afternoon, the obvious included a rather large shade tree which the groups of strollers continued to bypass or otherwise evade.

There was cause for the evasion, an obviously morbid cause. From one of the lower branches of the tree a corpse was hanging. From time to time it swayed stiffly. The cadaver had lately been a nondescript human male. From all visible evidence, he—it— was a suicide.

The approaching family head, who looked and walked more like a German than a Frenchman, paused to view the pitiful remains. Then he took out his pocketknife and with experienced dexterity cut the apparently self-rigged hanging rope. After the corpse plopped to earth, the man did not touch or seek to explain it.

Theodor Diesel, originally from Augsburg, Bavaria, had done what he considered his duty. He had acted to at least partly remove the shameful sight from public view; the next step was for the police or other public custodians.

Although a more or less hereditary mystic and intermittent student of metaphysics, leather craftsman T. Diesel did not offer any explanation to his nine-year-old son, Rudolf. Presumably the father took for granted that his only son was not old enough to comprehend such a relentless, overt fact of life as the self-imposition of death. Perhaps the leather craftsman from an expanding, complexly intermingled Germany had other reasons for not trying to explain the morbidly pathetic thing. Not too long before there had been at least one suicide in his own family. For worse or better, history quite often repeats itself. At least in the presence of his son, Theodor Diesel was not disposed to dwell on the dark and hellish topic of self-destruction.

History repeats, too, that life is hard; also that youth must harden itself to life, including the part called death. This acceptance is good German philosophy. Theodor Diesel's son, though born and being brought up in Paris, was German—good German with a proper German name, Rudolf Christian Karl Diesel— and never mind that silly Paris Sixth Precinct birth certificate which listed the young one's name as Rodolphe Chrétien Charles Diesel. This young Rudolf Diesel was a German, born and named for Theodor's own brother.

If the boy's mother had been along, some attempt at appraising or explaining the vagrant's suicide might have been unavoidable. But on this stroll father and son were alone—Frau Diesel and the two young Diesel daughters were not present. There would be no explaining or condoning suicidal tramps hanging from tree limbs.

But at very least here was a time and place for a practical lesson in surviving misfortunes. Evidently the bluing shape which now slumped on the ground had never learned that all-necessary lesson, all-necessary, at least, for a German.

Determinedly the father led on across another green and down a vale which was drained by a miry slough. As his son came alongside the purported drain ditch, the leather man—*maroquinier* was the Parisian word for his trade—turned about and quite deliberately tripped his son and watched the boy go sprawling into the foul mud.

No doubt a pleasure-soft Frenchman would term that rough treatment—or ring in some evasive French phrase meaning more or less the same. No doubt the surprised, seepage-soaked, mud-befouled nine-year-old, at least as of then and there, considered it rough treatment, too. But young Rudolf did not cry out. Because he was a German, the boy would one day understand. The trip-up was honestly in line of character strengthening, insuring against painful and inevitable upsets the son would surely encounter during the years ahead.

Upsets appeared to be the prevailing order—even in, or perhaps particularly, in Paris. For this Paris was far from being a happy place. She was being titillated by numberless minor upsets. There were indications, more than minor, that she would soon enough be wakened, not by silly ticklings but by welting lashes.

As he picked himself up and sought to wipe away the foul mud from his knees and jacket front, young Rudolf Diesel was not too happily aware of onlookers. There were the customary memberships of walking clubs and school clubs. There were the inevitable coveys of teen-age girls and boys, supposedly keeping with their own gender groups but making quite certain that girls and boys were never far apart. There were the hobbly old men and broadening Parisiennes who took on each other's company because there was nothing more satisfying for them to take on. There were also sprinklings of beggars and drunkards and trollops. And, of course, there were the lovers—young, old, medium, venal, cold, lusty, philosophic, and unassortable.

Inevitably there were wandering artists sketching or painting the Parisians at week-end strolls. *Maroquinier* Theodor Diesel made no pretense of being a "fine artist" in the sense of daubing water colors or oil paints or scratching with pencils or crayons. But in thinking, leather works, or book folios, in those areas T. Diesel considered himself an artist.

7

As a thinker, T. Diesel was a discouraged man. To him the Sunday strollers as a great group were symbols of this discouragement. Mostly they wore the dark, drab toggery of the bourgeois —the "safe and cautious people." There were multiplying reasons for Herrn Diesel to doubt that safety and caution could or would go on together. Perhaps the bluing corpse now slumped on the greening earth was symbolic of the impending rupture.

Eighteen years earlier, in 1849, Louis Napoleon had taken office as president of what was then called the French Republic. It did not long remain a republic or even a fluffy facsimile thereof. Similarly, Louis Napoleon did not long remain an elective head of state. By December, 1852, the so-called Prince President of France, who had been schooled in Diesel's home town of Augsburg, Bavaria, was formally declared Napoleon III, king-emperor of France. That climaxed a particularly black December for the Frenchman's dream of tending his vineyard and philosophizing in the sun or, weather directing, in the shade.

The passing of "Free France" was obvious as a country washday. In Paris and throughout most of the nation, the "imperial" names of streets, squares, public buildings, and resorts were being restored. Throughout Paris and much or most of *la belle France* the ever emotional but sometimes sincere hallmark phrases, beginning with *"Liberté, Égalité, Fraternité,"* were being erased, not only from statues and monuments but from living lips. In scores of town or village commons the Liberty Trees were being chopped down—pragmatically for fuel wood or burning fodder for the newfangled Vienna heaters. France had risen and stood forth as the *en garde* power of Europe. Paris had been reborn, not only as the capital of France, but both culturally and economically as the capital of western Europe. Now, in the 1860's, quite regardless of her grandeur and pre-eminence as mistress of *vivants,* Paris was no longer adequately cradling her own good people—much less other good and worthy Europeans, such as good Germans.

In his persistent use of "good German" for self-designation,

8

Theodor Diesel was actually speaking in metaphor. Technically he was Bavarian. Ancestrally he was a Swabian—Schwabisch— of the ancient tribe called Schwaben. Long designated as the "Scots of Germany," the Swabians were traditionally thrifty, astute, enterprising, and tough. Yet by the same broad designation they were also musical and strongly inclined to philosophy and poetry. The now federal state, then kingdom, of Bavaria was and still is a stronghold of the Schwaben. Through the centuries the town of Augsburg has been a center of the Schwaben; ethnically of the Württemberg Swabians, Augsburg is of Bavaria. But the Diesels were "political boundaries" Bavarians only. This one was a Swabian, self-placed, and from appearances misplaced, in Paris.

Gottlieb Theodor Diesel (who had arbitrarily dropped the "Gottlieb") had been born in Augsburg in 1830. He was German but markedly Swabian. His family name, "Diesel," first appeared 104 years earlier in the church records of Ludwigsburg, the summer capital of Württemberg or Schwaben; Stuttgart was the official capital. Earlier Diesel ancestors, presumably with Slavic surnames, had settled in the territory called Thuringia.

The picture-book land of the winding Saale River attracted the conquering Roman legions. It was a fertile land with mountaintops dotted with the mansions of agrarian nobility. In more recent times as small farmers took root in the land, many villages, including Pössneck, came into being. In the main, their people were characterized by oddly similar and umlauted names, including Tüssel, Tösel, Dössel, and Düssel. During the wretched years of the Thirty Years' War, a certain Hanns Tüssel settled in the hamlet called Pössneck.

In 1644, Hanns married sixteen-year-old Barbara Seisel, who died three years later after giving birth to their second son, Hanns. The younger Hanns grew up to be a butcher. He had to fight hard for recognition as a "master butcher," but eventually received the right to use that professional designation. In addition to owning his own shop, Hanns was the officially appointed appraiser of livestock for his village. His eldest son, Hanns Jörg

Tüssel, had a son, Hanns Christoff, who apparently renamed himself Hanns Christoff *Diesel*. Certainly his family name was to be so written in the church records in Ludwigsburg, where Hanns and his brother Nikolaus presently settled.

Ludwigsburg was a relatively new city, having been founded in 1704 when the Duke Eberhard Ludwig built a hunting castle there which he presently made his summer residence, and the town eventually became the summer capital of Württemberg. As such, it developed into a sizable place with many government officials. Nikolaus Augustin Diesel found a position there as a government office messenger. He had a son, Johann Christoph, who in 1740 became the first bookbinder and publisher in the free city of Memmingen, an ancient walled city south of Ulm.

At this point one begins to note the creative propensities of the line of Diesels. The son of the renowned Dr. Johann Georg Schelhorn urged Johann Christoph Diesel to publish a book of songs and verses written by the latter's father. Bookbinder Diesel accommodated and sent eighteen copies of the book to the city council in Speyer with a plea that the council buy them. But the book of sacred songs and religious thoughts and verse propounding the rapidly spreading Lutheran doctrine was not destined to be a best seller, certainly not in that still predominantly Roman Catholic community. Even so, it served to hallmark the name of Diesel as bookmen.

The second Johann Christoph Diesel followed in his father's footsteps. Honor came to him at the age of twenty-five when he was elected president of his bookmakers' guild. In 1845, Johann II died, leaving eight children and a detailed account of his life and the early lives of his offspring, with the births, christenings, marriages, and deaths carefully noted. One of his brothers (Johann Conrad) had acquired considerable wealth during his lifetime, but when he lost it all, he was unable to cope with his ill fortune and shot himself. This was the first recorded suicide in the Diesel family. It was not the last.

Theodor Diesel's father, the third Johann Christoph, was born in Memmingen in 1802. At eighteen, this Johann Diesel, who was lame, moved to Augsburg and became apprenticed to the bookbinder and publisher Johann Blosfeld.

Five months later he limped along to the larger Adenkofer shop in Munich. From there he went to Vienna. When he was unable to find work in the Austrian capital, the young Diesel with the badly crippled foot tramped the two hundred additional miles to Prague, thus demonstrating that Diesels were a determined line.

When the young traveler presently returned to Augsburg, he found himself a wife, young Sabine Riess, the daughter of an ironsmith. The young couple moved into a high-gabled, thatch-roofed house at Mauerberg C 117, in the midst of the old city's shopping district. Johann III set up a bookbindery on the street floor, directly below their lodgings. It was in this house on June 12, 1830, that their first son, Gottlieb Theodor Hermann, now the Parisian Diesel, was born. A second son, Rudolf, for whom Theodor named his only son, was also born in the living quarters above the bookshop.

At least intermittently Johann Christoph Diesel was capable of enterprise. In another shop on Karolinenstrasse he sold papers, pencils, and books. On finding the total returns insufficient to support his family, he began to sell other wares, such as portfolios and traveling bags. It was an era of severe trade restrictions, and leather tradesmen protested to the authorities who granted shopkeepers' franchises. The bookbinder was officially directed to stay with his bindery shelves.

By then Theodor Diesel's father was evidencing behavior patterns of which young Rudolf's were reminiscent. (The Swabian motto says not "like father, like son," but rather, "like grandfather, like grandson.") Johann III, to cite one instance, was deeply pacifistic. Even in 1866, when troops of the mighty Austrian armies marched through the winding cobblestone streets of Augsburg and the greatly feared Prussians came storming into position for the fierce, decisive battle of Königgrätz, Johann contended, and oddly enough, correctly, that the Prussians would not slaughter Bavarians. The fearful burghers felt differently. But for once the less than successful bookbinder foretold correctly. The Prussians, warriors that they were, simply did not hate the dream-dusted Bavarians. War does not just have to happen. To this both Johann's son and grandson would presently agree.

On Sundays old Johann would roam the spacious meadows by the gentle river with his butterfly net and drum; he was a prodigious collector of butterflies and caterpillars. He also searched the beautiful countryside for the appropriate, delicate foodstuffs for his many caterpillars. In the same whimsical, hobby-loving mood the pleasantly pixied bookbinder bought himself a handsome and prolific apple tree in an orchard. He made the harvest of its delicious fruit occasion for an annual and rather riotous picnic, attended by many relatives and a multitude of friends. The most acidulous of critics were disposed to admit that being crazy in a nice way could be rather good fun.

On an outing to Biburg in 1867, the sixty-five-year-old Johann Christoph Diesel picked a huge bouquet of heather. At home he suffered a fatal heart attack. The daintily flowered heather sprays from the Heide he loved so well decorated his plain wooden casket.

As the sadly confused son of a rather gaily confused father, Theodor Diesel found himself swept along by a profoundly confusing tide of history. In the beginning he was moderately sheltered. He went to the Protestant school in Augsburg. Then he became an apprentice in the shop of his father, learning the ancestral trade of bookbinding. That he wrought well was demonstrated by the exquisite leather binding of a Bible, his graduation work.

The era into which Theodor was being graduated, however, was not exquisite, or of rational handiwork. The rise of socialism, flamed in England by the impressive writings of the German philosophers Karl Marx and Friedrich Engels, spread rapidly through Europe. By the late 1840's the political revolution, born of onrushing industrialization, was sweeping over western Europe like a vast storm cloud. As a sensitive young man, Theodor felt deeply the crosscurrents of unrest. Three of his cousins expressed their reactions by emigrating to America. Theodor responded to the urge to seek bigger places. Paris was his particular choice.

The Augsburg which Theodor Diesel left in 1850 had barely 25,000 people. But it had an illustrious history. Its *Reichsstadt* dated back to 1276. Throughout the fifteenth and sixteenth centuries it had endured as a distinguished trade and cultural stronghold; in the sixteenth (1555) it was the scene of the great *Religionsfriede* between Lutherans and Roman Catholics.

The fourth generation of leather-working Diesels, therefore, left a renowned town for a more renowned city. After some sightseeing en route, Theodor and his younger brother, Rudolf, arrived in Paris and began job hunting. Their first Parisian employment was in a leather shop. But their father's hard times in the leather trades seemed to pursue them. Finding work only part of the time, the young brothers went hungry most of the time. Often their combined earnings were barely enough to pay for one adequate meal every second day. Otherwise, they frequented dingy wine shops where one could buy a small bottle of rich-bodied, dark red wine and then eat long, crisp loaves of bread for nothing.

Brother Rudolf, defeated and resigned, returned to Augsburg to scratch along as best he could. But Theodor stayed on in Paris. Apparently he never became what could be termed an integrated Parisian. Yet in his own moody and somewhat pedestrian estimate, Theodor Diesel was a kind of internationalist. He kept hoping that Paris might again be a truly international city. This hope was to be shared by the much better confirmed internationalist whom he married.

About the time the twenty-year-old bookbinder left Augsburg for Paris, twenty-three-year-old Elise Strobel was setting out from Nürnberg for London. Elise's family background was broadly similar to Theodor's, though more prosperous. Her father, Georg Friedrich Strobel, the son of a mirror maker of Fürth, owned one of the better shops in Augsburg. The shop sold exquisite crystal, excellent cutlery, and superior leather goods. Her mother was Kathrine von Schwer, whose family had fled to Nürnberg as a result of the Salzburg persecution of Protestants in the 1730's. It followed that the pretty young woman who was destined to be Rudolf Diesel's mother was the daughter of a prosperous merchant with a deeply religious wife.

After her mother's death in 1849, Elise Strobel served temporarily as homemaker for her seven younger brothers and sisters before deciding to set forth on her own. Her first choice of a special Baghdad was London. There she sought work for which experience had best fitted her, as a governess, and was successful.

In October, 1849, Elise wrote a long letter to her father, telling of the storied places she had visited in London—the museums and art galleries, the concert halls and theaters, and the opera where talented artists delighted her. She wrote, too, of the gravel-piled beach at Brighton and the gray-blue mysteriousness of the English Channel beyond. Throughout a five weeks' stay, Elise spent several hours each day watching the ever fascinating little sea. She wrote of Channel storms when violent waves pounded the shorelines and raised great towers of spray. It may have been premonition, for the same ever restless waters were to be the sea of destiny for her son, as yet unconceived.

Georg Strobel died suddenly in July, 1850. Elise hurried to Augsburg for the funeral. When Herrn Strobel's property was auctioned to settle the estate, Elise bought the combination dwelling and shop on the Carlstrasse. With the gentility typical of the "old South German," fellow Augsburgers refrained from outbidding the means of the heirs. While resuming her duties as caretaker for the younger Strobels, the eldest daughter began giving—more literally, selling—English lessons to the increasing throngs of local people who were interested in setting out for America. Between 1844 and 1850, German emigration to America had increased fivefold. Throughout a busy year Elise followed the tide by teaching English to additional prospective emigrants.

Then, for reasons not clearly explained, the practical young woman decided to leave her homeland again and try her luck in still another grand metropolis. This time she chose Paris. There Elise Strobel met Theodor Diesel. On September 10, 1855, after a brief courtship, Theodor and Elise were married.

They went to London to be married since Bavarian authorities would not grant permission for Bavarian nationals to be wedded in France. Indeed, somewhat comically, when the Diesels moved to Munich twenty-two years later, they were obliged to submit to a civil marriage ceremony in order to make legal in Germany their earlier marriage in England.

However, in September, 1855, when the new bride and groom returned to Paris and took a frugal apartment at 38 Rue de Notre Dame de Nazareth, they could not help observing that Paris was no longer the "wealth capital." By then Great Britain had sailed, banked, and politically maneuvered her way to being the foremost industrial and wealth-building power, a fact symbolized by Britain's first world's fair, the Great Exhibition of 1851.

By 1855, Britain, largely by way of England, was boasting an economy spined with steel. Even so, at the Great Exhibition the German Krupp *Werke* had shown a solid block of flawlessly poured steel weighing 4,300 pounds. The next largest block, displayed by a British smelter, weighed only 2,400 pounds. The German exhibits at the exposition were immensely impressive, although they lacked collective impact because the individual German states were still competing avidly among themselves.

At the time Theodor Diesel took himself a wife, Paris was playing glamorous hostess with its own world's exposition. For the time being England was preoccupied with the Crimean War. In lieu of the long, bloody siege of Sebastopol and the sagaed but ill-maneuvered charge of the Light Brigade (into the artillery death trap at Balaklava), France by way of Paris was learning mass showmanship and venturing into the sometimes glamorous industry of tourism.

Further, with Britain sunk into a decidedly ruinous war, Paris sought restoration of global eminence. Napoleon III moved with grim deftness to make himself the leader of western Europe and Paris its heart. The second Napoleonic France was not profoundly happy, but she was at least practicing an opportunism which, however unwise, was gaudy and bold.

As bona fide residents of Paris, Theodor and Elise Diesel strove to keep to the baffling pace. It was not easy. In one of the rooms of their modest rented apartment Theodor set up his leather shop. In the same apartment his daughter Louise was born in 1856, and then a first son, Rudolf Christian Karl, less than two years later, and after another eighteen months or so, another daughter, Emma.

The token population explosion called for larger quarters, which in turn called for a larger production of salable goods. Therefore, the Diesels moved to a two-story apartment on the Rue Gontaine au Roi, a location with an atelier for housing the leather shop.

There the French-designated *maroquinier* with the pervasive Swabian accent undertook a program of expansion, his least-fettered show of ambition. He employed as many as six workmen and two apprentices. As a matter of course, the three newly added Diesels grew up near the sounds and odors of a busy shop which worked from dawn till dark six days of every week. That, too, was thrifty Swabian and good Bavarian tradition. It would, or should, enable still another male Diesel to grow up to a worthy ancestral trade.

In due time, of course. There is no merit in having toddlers cluttering a workshop. Theodor Diesel listed his workshop as his "studio," and that without time or place for young children. Productive workmen were not to be disturbed. So young Rudolf was not permitted to bring any of his young friends into the shop. He was not encouraged to loaf there himself.

Rudolf responded in ways which were not invariably gratifying. He seldom played in the available yardway or with other children. Indeed, he seldom played. From toddling age the youngster had been unusually shy and strongly inclined to solitude. In solitude he frequently did strange and disturbing things; sometimes they were quite deplorable. For example, when left alone at home, little Rudolf once opened all the gas jets. Had it not been for the alertness and keen olfactory sense of one of the workmen, the entire establishment might have been blown to smithereens. Another time the odd and shy child took apart the much cherished family cuckoo clock and found himself wholly unable to reassemble it. That was doubly bad; he had destroyed a valuable property and instituted an undertaking which he failed to finish.

Theodor Diesel was frequently disturbed and repeatedly angered by his son's vagaries. At least he knew what to do: a leather craftsman knows how to swing a strap. When and as the more conventional punishments did not prove sufficient, Papa Theodor devised and applied special ones. One Sunday, he tied young Rudolf to a piece of furniture while the rest of the family took its regular outing. Again, when Rudolf told a lie, he was sent to school wearing around his neck a placard on which was lettered for all to see: "I AM A LIAR."

Despite the humiliating and sometimes brutal chastisement and despite his relatively poor French (the child spoke German quite well, and his linguist mother made a special point of teaching her children English), young Rudolf showed exceptional ability as a scholar. Much of the time he led his class at the Protestant school which he and his sisters attended. His drawing was excellent, almost unbelievably so for a youngster. The rest of his schoolwork was of such caliber that at twelve he was formally presented an inscribed bronze medal for superior scholarship from the *Société pour l'Instruction Élémentaire*.

What is one to do with such a boy?

There is evidence that as years hurried by, Theodor grew less certain he knew the answer. Himself a shy man, the *maroquinier* saw his only son as an exceptionally beautiful youth, a seemingly unpredictable liar, an occasional doer of malicious mischief, a frequent recluse, an exceptionally bright student, and a gifted artist. The boy looked like a poet, at times behaved like a poet. He seemed incredibly wise, yet oddly unworldly. He was not only handsome, he was beautiful; quite probably one of the most beautiful young males in Paris. Young Rudolf was a dreamer who dreamed in broad daylight. Yet, like the storybook bohemian, he was disposed to retreat to an attic or some remote corner or vacant hall where he was frequently, almost habitually, overtaken by the urge to make drawings or scribble in a notebook.

For good, or possibly bad, measure, the same young Rudolf Christian Karl was given to haunting museums. On a not distant avenue, the Boulevard Sebastopol, was the oldest technical museum in Paris and, quite probably, in the world. The place of hodgepodge was known as the *Conservatoire des Arts et Métiers*. Its building was the onetime abbey of Saint Martin in the Fields. It was a dingy and dankly ancient edifice crammed with dusty and oddly contrasting exhibits. These included all manner of tools, numerous models of early ships, and early models of steam engines. Each in its own way seemed to intrigue and attract young Rudolf.

But the boy's special delight and special magnet of interest was the original of the very first self-propelled vehicle known to history. This was a three-wheeled steam tractor, designed and built in 1770 by a certain Captain Nicolas Joseph Cugnot. It was substantially a round steam boiler, unprotected in front, mounted on great iron-shod wooden wheels. Hard as it is and then was to believe, almost a century before Rudolf Diesel was born, the contraption had actually run on its own power. On its first trip, so the guidebook tells, the thing had attained a top speed of about two miles an hour, somewhere near that of a tired ox. But, like the ox, it had to stop every few hundred feet to build up more steam.

Even so, the contraption invited use. Since in the 1770's the prevailing Versailles concept of usefulness was military, the French Army took over and set out to employ the self-powered steam contraption for hauling cannon. When so loaded and fired up, the mechanical primitive promptly went out of control and went clanging and roaring into a heavy wall. In the process of demolishing the wall, the vehicle was somewhat damaged, but

18

it remained approximately whole and quite memorable, at least to Rudolf Christian Karl Diesel. In the silent hallways the precocious schoolboy hovered to draw in his notebook pictures of the first steam tractor and various other musty but amazing exhibits.

In many ways their son Rudolf was "different." This became evident directly after his birth when Elise found herself unable to suckle the baby adequately. According to Swabian custom, the young parents gave the newborn one to a responsible wet nurse, who suckled and tended the man-child during the first nine months of his life, then returned little Rudolf in quite satisfactory condition to his parents.

Their handsome and undeniably gifted son was the Diesels' only problem child but by no means their only problem. Since practically all German tradesmen met serious difficulties in selling directly to Parisian markets, Theodor sold all the output of his shop to the German-founded retail store, Kellers of Solingen.

This action was motivated, not only by apparent Parisian prejudice against German-made goods (even when made in Paris), but also by Diesel's lack of capital to finance completely his own manufacturing enterprise. However, with diligence and considerable competence Theodor extended his manufactures beyond the beginning list of leather wares and added silk and velvet for pocketbooks and presently toys and games. These last items evidenced the Augsburger's notable but not much noticed flare for invention. He had invented a "light shedder" for placing over a burning gas flame; it long predated the gaslight cylinder later to become a standard lighting fixture.

Theodor worked hard but less than successfully. Financial straits lingered. He found it hard to meet his weekly payrolls. The seemingly inevitable slack seasons, usually two a year, grew more and more difficult to pull through. It was now that his wife, Elise, proved herself an exceptionally able business partner, besides being a very good homemaker. She helped with the shop and helped the shop to better standards of products and efficiency. As one example, she engineered the purchase and installation of a memorable new American invention called the Singer Sewing Machine, one of the first hundred produced.

In spite of the astuteness of Elise and the devoted long-houred labors of Theodor, their home factory did not flourish. Theodor Diesel, as a usually thin-pursed proprietor, was deeply if intuitively certain that the causes were far beyond the complexities of the leather and novelty trades. They related to the unstable position of France in the swaying and darkening fog of international affairs.

As the 1860's fumbled along, France's relations with the new confederation called Germany continued to worsen. This was obvious and, as Theodor saw and felt it, appallingly unnecessary. Regardless of broody and militant Prussia, Germany as a whole had no real quarrel with France, and vice versa. Bavaria and South Germany as a whole—for that matter, a plurality of the then thirty-nine German states—were inhabited by peace-loving peoples. As peace-abiding neighbors, they respected France, and most of them not already won as friends of France were certainly susceptible.

Yet the France of Napoleon III was doing little or nothing to be a friend to those "little Germanies." The fault here was definitely related to the self-advancing King-Emperor. And this, as T. Diesel saw it, was worse than ridiculous. After all, this Louis Napoleon personally knew the Germanies to the south. Actually, as already noted, a substantial part of the new Napoleon's education, to use a loose term loosely, had been in the Diesels' home town of Augsburg, traditionally a little city of friendly warmth.

The first Napoleon had beaten Prussia cruelly and overrun many of the German states. Understandably, few Germans had kindly feelings toward a France under reins of yet another Napoleon. The Bavarians tended to like Prussia even less than France. The Saxons, on the other hand, were, as they still are, the vacillating ones. Their traditional disposition was, and still is, to accept political advantages, or what seem to them political advantages, as they come. But the Prussians, at least a great many of them, dreamed of revenge against Napoleon, if not the real one, then the current and definitely obnoxious facsimile.

Apparently Napoleon III believed, or felt, or imagined otherwise. As Theodor Diesel saw him, "Pompous Louis" was anti-German in a most absurd and ill-advised way. This was more than the partisan belief of a worried Augsburg craftsman in Paris. As a monarch, Louis Napoleon was, on many fronts, being appraised as an ephemeral and crotchety head of state.

Even his American policies proved as much. At the outbreak of the Civil War, Napoleon III had sided dangerously with the Confederacy. Some of the fault here may have been attributable to incompetent advisers—"sorry intelligence" as the British were beginning to say. More, perhaps, was due to Louis' stupidity. The attempt to gain for Maximilian the "kingship" of alien Mexico bordered on the idiotic. Even the least informed of Diesel's hired help agreed that the Americas were no place for kings. Theodor Diesel agreed doubly; he was also a devoted reader of Jean Jacques Rousseau.

No phony kingdom, not even a foreign fabricated "empire," could possibly endure in America. And practically any intelligent onlooker could foresee that the rebelling Confederacy could not win the terrifyingly gory American Civil War. As Rousseau had ably stated, the Union forces had essential right on their side. They were fighting for man's liberation from literal slavery. The Union forces had also the advantage of resources, including American iron and inventions and German fighting men. Throughout the four earth-startling years of the Civil War, Germany's press had recorded with pride that entire divisions of the Union Army were manned in substantial or major part with *den Deutschen*; there were entire regiments and brigades in which German was the officially accepted language.

Following the Union victory, Louis Napoleon coldly abandoned his pawn Maximilian, yet pigheadedly evaded the obvious lessons of the American war. He continued to blow hot and cold with a foreign policy which no German could regard other than as grossly anti-German. He continued to repress the love of liberty.

Meanwhile, the anti-German sentiment in Paris continued to chill and annoy. The faltering leather and novelty shop continued to falter more and more tryingly.

Even so, Elise Diesel sought to keep on improving her competence as a mother, wife, and homemaker and to keep on hoping for the broader realization of her own small but worthy dreams of international friendships. As an adept language teacher, she continued to help her three children gain proficiency in the three languages she knew, German, English, and French.

At home, while dressing and breakfasting, Elise made a practice of speaking only English. The "trade language," including that of the occasional servants and most of the shop workmen, was German. At school, of course, the children learned and lived French. Conceivably the three young Diesels may have gained the impression that everybody in the world spoke his own language—different from all others; but at least the new crop of Diesels was growing up multi-lingually. And at least they were growing up with a considerable wealth of parental love. Elise Diesel adored her children; and Theodor, despite his paradoxical behavior, tried, with only occasional lapses, to be a good father.

This was evidenced quite graphically when the shop master succeeded in saving up enough francs to rent for one summer a "Sunday cottage" in the picturesque countryside of Vincennes. The retreat was fully as historic as it was idyllic. Charles V had built nearby a walled and moat-enclosed chateau. From the lively century on, French kings had used the wooded realm as a hunting site. The Diesels, of course, were more humble and temporary tenants, but for the three younger Diesels that summer was heaven on earth.

It had a few mundane limitations. The horse-bus fare from Paris to Vincennes was six sous a person. To reduce overhead, the Diesels regularly walked to and from their city home to their all too temporary country home. On Mondays the mother and her three young children took to the road at sunrise.

Although they were happy to do this, the opportunity was not to be repeated. The fortunes of the Diesel leather and novelty goods shop continued to dim as Germans grew less and less welcome in Paris.

Maroquinier Diesel continued to meet difficulties in collecting for his wares as well as in selling them. As a shy man he turned to his shy son with hope for a solution. Young Rudolf was appealing and winsome. Logically and reasonably he should make an effective collector. The greedy Parisiennes who owed for fancy purses and would not pay, even though they had the money, might take matronly pity on the shy and rather shabbily dressed boy who spoke reasonably convincing French.

Rudolf was appalled by the imposed chore of collecting. He dreaded it profoundly and evaded making the actual calls. One gathers he fibbed more than occasionally about them. The total experience added to his dread and fear of poverty.

In 1870 young Rudolf was to enter what we now call junior high school—*École Primaire Supérieure,* but the violent outbreak of the Franco-Prussian War intervened. As Bavarians, the Diesels were not actual enemies of France, but unhappily they were being regarded as such. By August the deplorable reverses of the French armies fanned the long-lingering anti-German sentiment into a spreading stain of hatred.

Paris by then was being crowded with *émigré* nationals. From a hundred nearby countrysides, farm families, with cattle, sheep, and portable household goods, came streaming into the capital. On August 28, 1870, the Imperial War Ministry published a decree that unless granted special permit by the *gouverneur* of Paris, all foreigners were to stand ready to leave within three days.

Plans were quickly formulated in the Diesel household. It seemed out of the question for the entire family to leave Paris. The father would go to London alone. The others would remain.

During the next two days Theodor went from one official to another to find out what papers the officials required to consider granting a stay, while the practical Elise began packing necessary articles. The extremely tense situation did not change appreciably until September 4, a Sunday.

On the evening of September 3, the Empress Eugenie was

notified of the battle at Sedan of two days before. That night General Palikao advised the hastily summoned cabinet of the capitulation of the army and their Emperor, Napoleon III. Next morning, noisy newspaper vendors shrieked the terrifying news. The Emperor was captured. The army of 83,000 Frenchmen had surrendered at Sedan.

That Sunday, aroused Parisians gathered everywhere in small, excited groups. Huge crowds milled in the squares, leaderless, bewildered, and afraid. Then a small but apparently organized group armed with crude weapons appeared and began to tear down and destroy all the many medallions and signs which showed the likeness of their unloved Emperor. Many of the better merchants of Paris who had earned plaques by catering to the royal household had displayed them proudly above their business establishments. Now the easily swayed street mobs seethed with anger and hatred of their fallen monarch. France, they vowed, was to be a republic again—for the third time.

To be on the safe side, the Diesels made ready to leave Paris at once, should a sudden need arise. They closed the shop and notified employees and suppliers. Provisions had to be made for an extended stay somewhere else in the event that further travel should become impossible. But the hard fact remained that there was no actual cash in the house. Business had been extremely bad for many months. Theodor set out anxiously to obtain a loan.

The next day the order for all foreigners to leave Paris was reissued. That day the Diesel family boarded a crowded train for Rouen, where another train stood by to carry the refugees to Dieppe. Very early next morning the jam-packed refugee train delivered its unhappy horde to a steamship bound across the Channel for neutral England. The overcrowded cabins were stifling, and the wet deck was miserably cold. Theodor Diesel managed to buy some bread and a bar of chocolate and so fed his family. At daybreak the Channel steamer finally cast off her lines.

After a miserable seven-hour trip across the wave-chopped

Channel, the Diesels arrived safely, but weary and seasick, in Newhaven. All five crowded into one cheap room at the London-Paris Hotel. They had a "passing" of hot tea and went to sleep. Two days later they reached London.

Leaving the three children sitting on one of the hard station benches, the parents went out into the city to look for such quarters as they could afford. After an hour they returned. They had found nothing. Now all five set forth to tramp the streets together, carrying their heavy luggage and looking for cheap lodging. At last, nearly exhausted, they found a simple two-room apartment at 20 Herbert Street in the Hoxton district. The three children slept on a couch and the parents in the bed. The small boxes they had brought along as luggage were used for chairs, table, and washstand.

The mother promptly attempted to find her old English friend of many years before and ask her help, but Miss Winton had moved away and could not be located. The parents were able to get in touch with the persons whose names their pastor had given them in Paris. In a short time the eldest daughter, Louise, then fourteen and, like her brother, quite brilliant, found a job as language and music teacher in a small private school. Theodor found work in a retail store at one pound a week, which was not nearly enough for his family to live on. Although after several weeks he found a slightly better-paying job, Theodor Diesel was faced relentlessly with the fact that he was no longer a shop master. He was just another head of an immigrant family, obliged somehow to scratch out a living.

Industrial Revolution

☕
☕

☕ AT LEAST FOR YOUNG RUDOLF the dark cloud had a
silver lining. He was able to enroll in a London school
and resume his studies. Even more enlightening, he had
an opportunity to become acquainted with the city that was
growing into an exciting and interesting world capital.

The young refugee paid frequent visits to both the magnificent
British Museum and the South Kensington Museum, which
promptly became his two favorite places. He also stood for hours
on the London Bridge watching with delighted fascination the
world-portraying traffic on the slow-flowing Thames.

News from Paris continued to be bad as far as France was
concerned. Two weeks after the Diesels had left Paris the German
armies began to besiege the French capital. Routine communications between the beleaguered city and the outside world were
shut off. However, some very ingenious devices were employed
to establish contact between the harassed Parisians and the outside world. Carrier pigeons brought messages written on thin
paper and inserted in metal capsules fastened to their tail feath-
ers. Notices printed on small sheets of paper were magnified
intensely by means of magic lanterns and thrown on a huge
white screen to be read by the populace.

Balloons—the first of which had ascended eighty-eight years
before—carried messages and, in some instances, men out of the
encircled city and across the Prussian lines at times when the

wind was favorable for the daring ascensions. In this way the "bitter ender," Léon Gambetta, escaped to fight again.

Young Rudolf Diesel heard provocative talk of and speculation about a steerable airship. As a matter of fact, a dirigible 125 feet long and 50 feet wide, to be driven by propellers turned by eight soldiers, was actually under construction in Paris, but it was not completely finished when the war ended. During the desperate breakthrough battles that November, an amazing dynamo developed by the Belgian Zénobe Théophile Gramme was used to floodlight from Montmartre, the highest point in Paris, a large combat area of Gennevillier.

Young Diesel was also learning a great deal not immediately related to the fierce, running war across the Seine, but all of real significance to a career which was to be highly important to most of mankind and its amazing composite which is termed civilization.

The great industrial expansion, which by now had shaped a world revolution, had begun in England, mainly in London, late in the eighteenth century. For approximately fifty years, until around 1820, the roots and smoky fruits were primarily English. During the ensuing four decades the Industrial Revolution had spread substantially over the European continent and strongly into the United States. In Germany, industrial developments had moved along with even greater rapidity than in neighboring France.

In 1819 the King of Prussia removed the internal customs barriers and made the entire Prussian state one fiscal unit, thus opening the door to much greater freedom of commerce. Each of the other thirty-eight independent German states, however, still had its own customs barrier. Individually and combinedly, these barriers had all but halted interstate trade. In 1812, for example, on the 125-mile trip from Fulda to Altenburg, a traveler crossed thirty-four boundary lines and went through the dominions of nine sovereign monarchs. Then, in 1834, a *Zollverein*, or customs union, was established which embraced seventeen

German states with twenty-three million people. It simplified trade considerably. By 1842 this system had been extended to nearly all of the thirty-nine German states. This extension resulted in a tremendous boost of internal trade and the beginning of many great industrial empires.

Many of the industrial giants of the twentieth century had their modest beginnings during the early nineteenth century. To cite one instance, in 1810, Friedrich Krupp purchased a small forge at Essen and worked on the problem of manufacturing cast steel. Carried on by the son after the founder's death, the Krupp steelworks soon reached world prominence. Similarly, in 1837, Emanuel Borsig founded a machine shop in Berlin which eventually grew into the vast locomotive works there.

In 1847, Werner Siemens opened a repair shop for telegraphy apparatus in the back yard of a Berlin apartment house. A year later he was awarded a government contract to build a telegraph line from Berlin to Frankfurt. Then he invented the modern dynamo which converts mechanical energy into electrical energy. In 1879, Siemens demonstrated the first electric railway engine. The mighty Siemens-Schuckert and Siemens-Halske *Werke* grew from these early inventions. From these and many other instances young Rudolf Diesel learned in London the memorable industrial landmarks of his future homeland.

In the five-year period following 1850, consumption of raw cotton in Germany tripled; between 1848 and 1857, production of iron ore trebled. With the perfection of the Bessemer process in 1856, the per capita pig-iron consumption quadrupled in the ensuing fifteen years. Steam power was being used extensively.

The development of the steam engine was, of course, of major importance in the growth of industry in Germany and practically everywhere else. Although the Scotsman James Watt is generally, and rightfully, credited with the first practical steam engine, the first practical application of the principle actually goes back to about 100 B.C. It was then that Hero of Alexandria used a crudely constructed steam-powered device to open and close temple doors for the amusement of his amazed royal guests.

The Italian Giovanni Branca and the Englishman Somerset again worked with this principle in the seventeenth century. In 1690, the Frenchman Denis Papin, a professor of physics in the University of Marburg and curator of the Royal Society, invented a digester engine which employed a moving piston in a cylinder. Soon thereafter, Thomas Savery devised a boiler in which steam was produced constantly, eliminating the cooling and heating of the cylinder. However, Savery did not employ a piston at all. His invention was an atmospherically propelled pumping device.

About 1705, Thomas Newcomen successfully combined the Papin piston engine and the Savery boiler. An automatic device for the control of the steam valves was suggested by a lazy boy who tended the engine, purely to serve his personal, slothful purpose.

James Watt made mathematical instruments at the University of Glasgow. While repairing a Newcomen steam engine, he invented the means of forcing the piston back and forth in the cylinder by simply closing both ends and applying steam. This, and the separate condensing chamber, made the steam engine of Watt's design a practical machine.

Although patented in 1769, the steam engine was not a commercial success until several years later when Matthew Boulton became a partner in the manufacturing enterprise. In 1785 a steam engine was first used to power spinning machinery in the mills of Nottinghamshire. With the perfection of the Watt-Boulton steam engine for industrial use, inventors worked feverishly to adapt this new prime mover to transportation.

Constant improvements of navigable rivers and the building of canals provided a basis for a vast transportation system which actually dated back to the ancient days of the Babylonians. Tramways in Germany dated back to the fifteenth century. Heavy, squared timbers, laid parallel on the ground, the distance between them corresponding to the gauge of the vehicle, were used as a track. Later, iron strips were placed on top of the wood to afford protection against excessive wear. Eventually, iron rails supplanted the earlier forms.

Public coaches were pulled by oxen or horses when William Heldey, Richard Trevithick, and George Stephenson worked on adaptations of the steam engine to locomotive power. In 1825, eleven years after the completion of the first steam locomotive, the first straight-line English railroad was opened between Stockton and Darlington. The short train reached a speed of up to fifteen miles an hour. Five years later, Stephenson's famous *Rocket* reached a speed of twenty-nine miles an hour. The first train pulled by a steam engine in the United States ran from Albany to Schenectady in 1831.

The first railroad line in Germany was opened in 1835 on the four-mile run from Nürnberg to Fürth. The important cities of Dresden and Leipzig were connected four years later by a rail line. By 1850, railroad-building activities covered 3,633 miles of trackage to mark Germany the great pioneer railroad nation. In Prussia alone, 290 miles of track had been constructed by 1840, and twenty years later the mileage had increased to 6,890. This fact, too, was to influence the destiny of young Rudolf Diesel.

In 1690, the French physicist Denis Papin had suggested that his steam engine could propel boats. In 1707, when he demonstrated successfully a model boat on the Fulda River at Kassel (Germany), angry river boatmen, fearful of losing their livelihood, destroyed his model and threatened his life.

In 1786, a Virginia mechanic, James Rumsey, had built a steamboat and demonstrated it on the Potomac River, achieving a speed of four miles an hour traveling against the mild current. Four years later, the American John Fitch carried passengers on his propeller-driven steamboats for several months. At the same time, William Symington built a steamboat in England. In 1807, Robert Fulton built the famous *Clermont* and made a trip up the Hudson River from New York to Albany, a distance of 150 miles, in thirty-two hours. His boat with its Watt-Boulton engine was unquestionably the first commercially successful steamboat, and Fulton, a genius of a promoter as well as an inventor, led in popularizing the "new" means of transportation. Fate was to permit young Rudolf Diesel to make the steamboat efficient from the standpoint of power use.

In 1819, the steamboat *Savannah,* equipped with auxiliary sails, made her first Atlantic crossing in twenty-five exciting days. The *Curaçao,* the first European ship to use steam alone, crossed the Atlantic from Antwerp to Paramaribo, Dutch Guiana, in 1827 in about the same run time. When, in April, 1838, the *Sirius* crossed the ocean in only eighteen days, the reign of the sailing vessels was definitely threatened. In the same month the magnificent *Great Western* needed but fifteen days, and thirty-six tons of coal daily, to complete the Atlantic trip. This extraordinary ship and its historic feat contributed notably to the founding, in 1839, of the Cunard Steamship Line.

While records indicate that iron scows were built by the English ironmaster Wilkinson as early as 1787, it was not until after the advent of the steamships that Archimedes' ancient law of floating bodies was recalled and ships of iron were constructed. On land as well as sea, iron and other great base metals were already being marked as materials of progress, prosperity, and national survival.

The Diesels' final year in Paris had seen the completion of the Suez Canal, which revitalized British and Scandinavian shipping and helped further to underwrite Britain's prime place as a commercial sea power. For good measure (there were some who still said bad measure), Frenchman de Lesseps' amazing venture in promoting a man-made water route to tie Europe, Britain, and Scandinavia more closely to the Indian Ocean and the fabled Orient was another great boon to steam-powered ships.

In London, young Rudolf Diesel had opportunity to hear much and learn a little about the power available from electricity. French and British schoolbooks alike recounted Dr. Benjamin Franklin's demonstration of the identical properties of lightning and man-generated electricity. Schoolbooks were beginning to tell, too, of German physicist Johann Thomas Seebeck's discovery of the theory of the magnetic field and of France's great André Marie Ampère, who was busy proving the basic law of electrodynamics.

England's Michael Faraday, builder of the first electric dynamo, was not yet famous, nor was Germany's Heinrich Rudolf Hertz, who would presently establish in workable principle his theory of Hertzian waves to link together the common elements of heat, light, and sound. Thomas Alva Edison was then twenty-three years old and more than a fledgling inventor. The German-born Charles Steinmetz, destined to become a superior electrical engineer, college teacher, author, and inventor, was, like Rudolf Diesel, among the living and growing, although in 1870 he was only five years old.

In the eventually decisive field of communications, the telegraphing of a message from Munich to Bogenhausen by Carl Steinheil in 1837 was the first step toward encircling the globe by electrical impulses. That same year Sir Charles Wheatstone telegraphed a message from Euston to Camden, England. But commercial telegraphy waited until May 24, 1844, when Samuel Morse dispatched by wire-carried electrical impulse from the United States Supreme Court chambers in Washington, D.C., to a receiving point in Baltimore, Maryland, the words "What hath God wrought."

During the year preceding Rudolf Diesel's birth, American Cyrus West Field began his ordeal of laying an under-ocean cable from Ireland to Newfoundland. The first venture failed. While Rudolf Diesel was being wet-nursed, capitalist Field's two cable ships started from the mid-Atlantic, each with its weighty cargo of ocean-bed cable. After three breaks, the *Niagara* reached Newfoundland and the *Agamemnon* arrived at Valentia Harbor, Ireland, with their respective ends of the first Atlantic cable. President James Buchanan and Queen Victoria exchanged greetings by under-ocean electronics. But Rudolf was seven before cable communication between the Old World and the New was successfully established.

More directly related to the interests and destinies of the shy young museum haunter was the rather vague birth and infancy of the internal-combustion engine. Its first inventor, at least in principle, was quite probably the Dutch astronomer Christian Huygens, who came up with an "inner power" engine in 1680. Huygens applied a principle advanced three years earlier by Jean de Hautefeuille, inventor of a device for drawing water. Huygens' machine used a gunpowder charge placed in an enclosed cylinder. A vacuum was created within the cylinder when the exploded gases cooled; the piston moved by the external atmospheric pressure rather than by pressure created by the explosion. While the cumbersome mechanism worked, it was far from being commercially practical.

During Theodor Diesel's boyhood the Englishman William Barnett obtained a patent for a cylindrical device for compressing an ignitable vapor charge and shared with a fellow Englishman, Douglas Clerk, the concept of the engine with a two-stroke cycle.

When Rudolf Diesel was four, the French engineer, Jean Joseph Étienne Lenoir constructed the first practical internal-combustion engine, using illuminating gas as fuel. His compatriot, Alphonse Beau de Rochas, is generally credited with having stated the principles of the four-stroke-cycle engine. But the first really practical internal-combustion engine was still awaiting inventors. They turned out to be the Germans Nikolaus August Otto and Eugen Langen. By then (1876) Rudolf Diesel was himself a respected German inventor.

For the present, however, the shy but perceptive son of the displaced leather craftsman had opportunity in London to glimpse more than casually the earth-changing phenomenon called the Industrial Revolution, for both in productive fact and by purse strings, London was its fast-growing capital.

The smoky gray lines of factories, many of them deplorable firetraps as well as sweatshops, were dotting the city's outskirts and being built in suburban communities and many other cities and towns of England. Very few of the mills were beautiful; great numbers were needlessly and obnoxiously ugly. As yet industry had established no convincing correlation with beauty or artistic expression.

Young Rudolf's first impressions reflected surprising maturity. On the whole, the English factories seemed to him distinctly abominable places to work; their tenements or "labor quarters" were fully as uninviting. The foul smells of coal smoke mingled with the more noxious odors of filth, of both humans and machines. For the most part, as the twelve-year-old noted with a marked degree of concern, the shop buildings were needlessly lacking in lighting and ventilation.

Work hours were even longer than those his father kept; from foggy dawn until long past dark. As the German-descended Queen Victoria pointed out, "Labor should be worthy of its hire." In the main, labor was more than worthy, but many of its hirers were unworthy of their hirelings.

To an appalling degree, Britain's industrial "expansion" was powered and sustained by child labor. Tens of thousands of its wage slaves were younger, certainly no older, than Rudolf Diesel was. Reportedly there were "hard kept" mills and "soft kept" mills, but Rudolf heard very little specific description of the "soft kept." As he later recalled, he never had a chance to take a firsthand look at one.

In human terms the English mills were needlessly cruel as well as stupidly greedy. In general their adult labor was hard driven and very badly paid. Abuse of child labor, however, was the self-perpetrating and wholly unpardonable sin of English mill management. By tens and hundreds of thousands English children were going to factories instead of to schools. As factory workers they were chattels. Floggings were more or less routine. Rudolf Diesel saw boys of his own age with prison pallor and bleeding welts.

Industrialization was making parts and parcels of England rich and smug. It was making greater parts of England degradingly poor. A new British peerage was rising from the muck of its ruthless, conniving betrayal not only of England but of human decency everywhere.

The foregoing, at least, was the understandably scrambled point of view of uprooted Theodor Diesel. But the twelve-year-old son of the unhappy *émigré* sensed another aspect of London's remarkable condition which was being called industrialism. Although most of the mills and their environs were ugly, many of the new machines were beautiful to young Rudolf, at least by virtue of function. They "ran and did," and in this aspect the young refugee sensed both beauty and creative artistry. England had poets who spun beautiful verses. And machines which wove beautiful cloth.

But for the time England had scant place for unemployed tradesmen. Try as they did, the Diesels found making ends meet virtually impossible. Theodor and Elise thought longingly of a return to Augsburg.

Theodor's younger brother, Rudolf, advised against the move. Jobs were also scarce at home. Business in Bavaria was grievously disturbed by the war. The wiser course, as Brother Rudolf saw it, was for the renegade Diesels to stay on in London and send their son to Augsburg to study. After all, with Theodor and daughter Louise both employed, even at stingy pay, the family could at least eat. An obliging husband of a gracious cousin of the Diesel brothers, Professor Christoph Barnickel, was willing to keep young Rudolf in his home and care for him. This would lift a sizable share of Theodor's burden and at the same time provide his son a safe haven, a good school, and a token reunion with lovable old Augsburg. The boy could stay at least for the duration of the war and, or until, the fortunes of Theodor improved.

Theodor and Elise accepted the offer and wrote Christoph and his wife, Cousin Betty, that they would send their son along as soon as his winter clothes could be obtained from Paris.

Apparently the clothes were recovered quite promptly, for on the first day of November Rudolf boarded a steamship at Harwich, and set forth for Rotterdam. He wore a collar placard giving his name and the name and address of nearest kin in Germany, Uncle Rudolf. He carried one light satchel of clothing and an oversized lunch basket with food enough to last several days. The latter was most provident. The trip required eight tumultuous days. But it terminated safely at the Schäzlerstrasse 8 and a breakfast board crowded with welcoming relatives.

Rudolf wrote to his parents the day after his arrival in careful detail about the eventful trip. Bad weather had delayed the steamer crossing the Channel for over seven hours. His traveling box had been severely damaged, but he had managed to bring it safely to Augsburg. In Rotterdam he had taken a train to Emmerich, where he ate and slept. The following day he had boarded a train for Cologne and there changed to another one for Frankfurt.

The train had been considerably delayed because of heavy military traffic, and he had found it necessary to stay overnight in an inn in the city, sharing the room with another traveler. Since his companion was a total stranger, and perhaps untrustworthy, Rudolf had thoughtfully left his silver watch and small amount of change in the safety of the care of the innkeeper.

Next day he had taken an express train to Würzburg. There he had changed to the combination passenger-and-freight night train, which had stopped for the usual half-hour waits at every town and hamlet along the road. The coach in which he rode had been hooked to different trains several times, and he had been unable to sleep. When he got extremely uncomfortable, because of the severe cold during the long winter night, he had taken a glass of strong schnapps. The alcohol in the drink had warmed his insides pleasantly. But thereafter and during the remainder of the journey he had suffered from a sore throat, a toothache, and a painful earache.

Rudolf told of his affectionate reception by his relatives and mentioned how homesick he was. He wondered how his sister Emma managed shining the shoes and building the fire. And how was Louise, he asked, and did father still have work or a prospect of earning more money?

The Barnickels kept the ailing and exhausted boy at home for several days before sending him to school. The professor observed that the youngster was a well-behaved "dear boy," to whom anyone could immediately lose his heart. Young Rudolf, as described by his professor cousin, was modest and spoke intelligently, more like a mature adult than a twelve-year-old boy.

Song of the Homeland

RUDOLF HAD SPENT barely eight weeks in London. As a kindlier fate directed, he was to spend five studious and comparatively happy years in Augsburg, his ancestral home.

These years were happy as well as fruitful. Always an intriguing town, old Augsburg was gradually rousing to a somewhat drowsy leadership in a distinctly South German version of industrialism.

Old Augsburg is truly old. When Caesar's legions poured over the Germani, it was even then ancient, a settlement at the crossroads of northern Europe, Italy, and the Levant. The Romans called it Augusta Vindelicorum.

By 1870, it had gradually taken its place as a noteworthy commercial city of around fifty-thousand people. The change-over had begun nearly half a century earlier—in 1824 when the steam engine made its entry. Augsburg had ever been a river town. Its location at the confluence of the Lech and Wertach rivers made relatively easy the building of canals which checkerboarded and clearly bisected the town and permitted the building and effective operation of many water wheels. This gave Augsburg a special advantage as a weavers' town. For centuries it had led its area in the manufacturing of linen, cotton, and other cloth. The advent of the steam engine had allowed widespread expansion of the textile industry.

Clothmaking of course, is an ancient art and trade. Yet between the time of Homer and the American Revolution, there had been only three decisive advances in weaving. In the fifteenth century the spinning wheel was invented. In 1767, James Hargreaves invented the jenny which operated several spindles from the same power source—in time as many as eighty spindles. Then in 1784, an English clergyman, Edward Cartwright, invented and built a successful power loom. Next, the able Yankee tinkerer and teacher, Eli Whitney, devised the cotton gin, with which one worker could "clean"—remove seeds and burs—as many as three hundred pounds of cotton in a single day instead of the one pound a day averaged by combing.

But all the historic textile machinery was powered by people— with strong arms and backs. The coming of the steam engine changed this. Old Augsburg was among the pioneers in making profitable use of the stationary steam engine; its cloth mills were by no means the only industries benefited.

In 1824, inventor Koenig himself had installed, on order of the Augsburg printing shop of Freiherr von Cotta, an English-built steam engine for powering the firm's first speed printing press. There were immediate and profound social and aesthetic complications. The steam engine was appallingly noisy. It huffed and roared like a monstrous, ever belligerent dragon.

Publisher Herr Stegmann had offices directly above the uproarious iron dragon whose racket he endured stalwartly. But the gentle Herr Wiedemann, second in charge of the printing enterprise, preferred to work elsewhere. The hired man, who had served faithfully for over ten years, gave notice. He valued his life, sanity, and eardrums—all the more so since he had a wife and several children to support. A prominent elderly merchant refused even to walk the street on which the plant was located. Fearful nuns in a cloister directly behind the printing plant unleashed a storm of protest against the owner. But the new, frightening, and terribly noisy steam engine remained.

Augsburg was developing into a notable industrial city. To clinch the evolution, the banker, tobacco manufacturer, and textile factory owner, Ludwig Sander, established a machine

39

factory in 1840. In 1844, a nephew of inventor Koenig, Carl Ludwig Reichenbach, the machine engineer of Freiherr von Cotta, took over the factory in partnership with his brother-in-law, Carl Buz. They renamed it the C. Reichenbach *Maschinenfabrik* and began building steam engines, water wheels, and various other machinery. By 1870, when the young Rudolf Diesel first came to Augsburg, the *Maschinenfabrik* employed more than six hundred men and roared along as the city's prime symbol of the new industrialization.

One of the more immediate aftermaths was the rapidly growing need for adequately schooled power engineers. Here young Rudolf Diesel saw a unique personal beacon light. Prior to 1870, two technical schools had been established in Bavaria after the pattern laid down by Napoleon I and followed successfully in France. In 1833, a third polytechnic school had been established in Augsburg. In 1864, the three institutions were combined into the Munich *Polytechnikum,* later the *Technische Hochschule* (Technical High School).

At the time of young Rudolf's arrival the Augsburg school was closed except for the mechanical shops, but a preparatory technical school had lately been established to supplement the existing facilities. Before Rudolf Diesel could enter this school, he had to attend the *Königliche Kreis-Gewerbeschule* (Royal County Trade School) for a three-year period. This school was located upstairs in the former St. Catherine Cloister, which housed an art gallery downstairs. Paintings by Hans Holbein the Elder, Hans Burgkmair, Cranach the Elder, Tintoretto, Van Dyck, Dürer, and Leonardo da Vinci hung there.

The more than seven hundred paintings were from suppressed churches and convents and from other old galleries. Great names and great history lived on in the bluish shadows of Augsburg.

The ready affinity of "pure art" and mechanical arts was thereby demonstrated with a true and noble forthrightness. There is good evidence that young Rudolf was already inclined favorably toward great art. There is good evidence, too, that his appreciation was shaped and sharpened by this humbly magnificent gallery on the floor below the school. Occasionally, also, probably not often enough, Rudolf played in the streets and near the old wall gates where Martin Luther had walked nearly 350 years before.

Although admittedly preparatory, the Royal County Trade School was a technical school. Various machines were displayed and used for demonstrations in a large classroom. There was an excellently equipped chemical laboratory and regular workshop with lathes and hearth and all manner of mechanical devices. Rudolf enjoyed the conventional textbook studies as well as the practical schooling, standing tenth in his class at the end of his first semester there.

Meanwhile, the Franco-Prussian War raged on. Paris held out stubbornly. Parisians were starving, but they refused to surrender. The Diesels in London had hoped to return to Paris after the war. Now their hope had changed to a desire that they could go back to Augsburg. Maybe they could open a store there. For reasons not clearly stated, the Augsburg relatives again advised against such a move. Theodor Diesel, his wife, and their two daughters remained in London and struggled on as best they could.

Word came to them of the capitulation of Rouen in December and then on January 28 of the fall of Paris. When the armistice was signed a month later, hopes for an early return to the French capital rose considerably. By July, Theodor and Elise and their two daughters responded to impulse and returned to Paris. They were shabby poor, but they had the courage to begin again. Relatives approving, young Rudolf stayed on in Augsburg.

His frequent letters to his parents were full of homesickness, although his foster parents were unusually kind to the now thirteen-year-old. On the surface, the homesickness was not difficult to diagnose. Rudolf had no close friends in school. His parents were unable to send money for new clothes or even to pay his bare living expenses. Rudolf decided to give mathematics lessons to other students and earn a little money that way. Since he was not a German citizen, he was ruled ineligible to receive any scholarship assistance, but Superintendent Pfeiffer somehow succeeded in raising sixty gulden for the winsome and studious boy who temporarily had neither a home nor, legally speaking, a country.

On March 24, 1872, Rudolf Diesel was confirmed in the Lutheran church. The certificate was made out to Rudolf Christoph Johann Diesel, instead of Rudolf Christian Karl. It seems likely that either the boy or his relatives gave the name incorrectly. There were no ruinous aftermaths. The Bible verse printed on the certificate (Genesis 12:1) seemed touchingly appropriate: "Go from your country and your kindred and your father's house to the land which I will show you." The following verse, "And I will bless you, and make your name great, so that you will be a blessing," was certainly prophetic.

Six days before his confirmation, on his fourteenth birthday, Rudolf decided to become an engineer. He wrote his parents of his decision. He also vowed to become first in his school class.

By the end of the semester, in August, Rudolf was fifth in his class. Meanwhile, he continued to prove himself a deep and conscientious thinker. Confidential conversations usually centered on his future. He was greatly concerned about his formal citizenship as an adult. In this connection he urged his father not to become a citizen of France, for in that event Rudolf, still a minor, would automatically be a French citizen. He felt himself a German. He also doubted that he could win a scholarship if he were a legal citizen of France. Besides, then as now, military service had to be considered. Compulsory military service in France was nine years; in that era nine years was half of a man's average working life. Rudolf hoped to begin work as a mechanical engineer at the age of twenty-one. Even the required three years of military duty in Germany would gravely interfere with his plans.

While pondering the future, he found at least a few opportunities to live for the present. On one happy occasion, in company with an uncle and a cousin, Rudolf spent a few days at Lake Starnberg. In a letter to his parents he described the superlative Alpine scenery and the breath-taking beauty of the lake. After one of the two steamers on the lake nearly swamped his small rowboat, the boy watched admiringly as the skillful fishermen handled their craft in treacherous waves raised by passing steamers. The wide-eyed visitor was given permission to try his hand at steering a boat. He succeeded quite well. In celebration he carved his initials into the soft bark of a young tree trunk. The thought of perpetuating his name apparently had crossed his young mind.

Rudolf Diesel took his final examination at the trade school at the age of fifteen. By then, as he had vowed, he was first in his class. But his next step was worrisome. His father had still not done anything definite about renewing his German citizenship. Rudolf's certification of German residence was little more than a passport visa. He would be obliged to wait at least five years before he could himself seek formal German citizenship. Meanwhile he feared he would be unable to attend the Polytechnic High School in Munich without a scholarship. And if he could not study at the school, he could not become a mechanical engineer.

His parents opposed additional schooling. In their lives money had always been scarce, and by letter they strongly urged the boy to abandon his plans for a higher education in favor of a paying job. They pointed out to him that manual labor was one of the proudest tasks. With straightforward logic, Rudolf replied that, if he left school now, he would have to work for three years as an apprentice with no wages before he could earn anything at all. If, on the other hand, he finished his studies at the technical school (although *Hochschule* is literally translated as "high school," the level of course work is equivalent to that in a college or university), he would be well qualified for profitable employment. The parents agreed to discuss the matter in person. Thus, after three years' absence, Rudolf took a train to Paris for a reunion with his family. He rode on the cheapest train, a single passenger coach attached to a local freight; the journey of barely three hundred miles took forty-two hours.

The reunion in Paris was less than happy. The Diesels now lived very meagerly in a tenement at 127 Boulevard Voltaire. The neighborhood was one which had seen hard times. Rudolf realized that he could never again expect financial aid from his parents. They simply had nothing. Theodor had attempted valiantly to establish another leather shop. The returns were mostly in the prestige of being a shopkeeper; actual earnings were less than those of common laborers in a badly depressed area.

Seven weeks after Rudolf arrived in Paris, his older sister died suddenly, possibly of heart failure. The family was understandably shocked and grief-stricken. Louise was talented as well as loved. On the brink of womanhood at sixteen, she was a gifted musician and for more than three years had helped the family purse by giving private piano lessons.

Rudolf was deeply shaken. A more comforting reaction appears to have been his resolution to learn to play the piano, thereby in some small part to carry on his sister's great gift and devotion.

Following Louise's death, both parents grew increasingly interested in the supernatural. Indeed, Theodor had long been drawn toward metaphysics and deeply submerged in mysticism. The grieving parents attended a spiritual séance, hoping to establish communication with their beloved child. They came away convinced that their departed daughter had communicated with them, through a medium, indicating that she was very unhappy in the world beyond. This experience added to their already heavy burden of grief. Elise did not abandon her faith as a good Lutheran, but Theodor became an ardent spiritualist and seemed to be breaking church ties.

After five unhappy months in Paris, during which his parents showed evidence of premature age and anticipation of abject poverty, Rudolf was thankful and greatly relieved when Herr Professor Barnickel and his good wife Betty renewed their invitation and reiterated that the "dear, sweet boy" was now part of their home. On October 1, 1873, when the new industrial school reopened, Rudolf proudly enrolled in the mechanical-technical division to study mechanical engineering.

44

Physics, mathematics, and mechanical drawing were almost instantly his favorite studies. Living with the Barnickels was even more pleasurable than before. Four sons of a wealthy widower named von Stetten had also been chosen to live with the professor. This time the place of residence was a spacious near-mansion owned by banker von Stetten. Here Rudolf had his own comfortable room, which faced a beautifully kept flower garden.

While it was a most enjoyable change from the bleak little apartment in Paris, the school itself was Rudolf's great joy. In the tall glass cabinet in the physics laboratory was a pneumatic fire apparatus. It was built like a bicycle pump. The barrel was made of glass, and one could watch the process of the compressed air igniting a spark at the end of the cylinder when the plunger was pushed down forcibly. What effect this *Feuerzeug* had on the thinking of Rudolf Diesel and on his later invention is not surely known, but the basic idea of combustion procedure for what presently would be the Diesel engine was here practically demonstrated.

After passing his semester examination with excellent grades, Rudolf again visited his parents in Paris. Again, his mother suggested that the son become a good mechanic by practical work rather than by the study of "theoretical problems." The financial condition of the older Diesels had not improved. Elise was still shaken with grief. Theodor mooned about in a strange, purposeless daze.

Back once more in Augsburg, Rudolf was able to secure from the considerate Superintendent Pfeiffer a stipend of four hundred gulden or guilders—about $165—which would insure his entrance into the high school for the coming semester. When he was graduated from the trade school, he was the youngest student ever to pass the final examination; indeed, his grades were the best ever recorded in that school—"excellent" in eleven of the twelve subjects. Herr Professor Barnickel's predictions regarding his young cousin were handsomely fulfilled. At eighteen, Rudolf Diesel had proved himself an outstanding scholar.

The Bavarian educational director, Karl Max von Bauern-feind, who came to Augsburg to interview personally outstanding students for scholarships at Munich, examined Rudolf Diesel. In recognition of Rudolf's brilliant record, he promised him two scholarships of five hundred gulden yearly, abundant for meeting all expenses at the Polytechnic High School in Munich.

During the school year, young Diesel earned additional cash by teaching French to fellow students, using the money to pay his room rent and to rent a piano. He was determined to learn more of music. Otherwise, he lived most frugally. For lunch, his main meal, he rarely spent more than fifty pfennigs (about twelve cents). He did not grow fat, but he did save money and, quite probably, he shared his savings with his hard-pressed parents.

It was a happy school in a gay city, but Rudolf was not there for gaity or even much sociability. His closest school friend was Friedrich Lang, son of a nursery owner and of modest circum-stances. The two students rented a room on Barerstrasse. Rudolf promptly settled to his studies. With quiet determination he attended all of the lectures at the school, studied English, and practiced the piano daily. Occasional long walks provided his recreation. He had no interest in the frivolities indulged by some of the other students. Scholar Diesel had set himself a high goal in life and was determined to reach it.

Following 1876, Rudolf wrote his sister Emma, whom he apparently regarded by then as his closest of kin: "The dear God has not withheld the help for which I have prayed so insistently, but also blessed me in immeasurable amount, so that I attained the highest mark in every subject, a result for which I had not dared to hope and for which I am thankful from the depths of my heart." He added in the same letter: "But the matter also has a material side, and since we are on this earth, on which food and drink, clothes and shoes are a consideration, one has to speak of this side too."

Rudolf's rapidly developing attraction to mathematics ap-peared to re-arouse his artistic talents. He illustrated geometry exercises with clever pen drawings. Presently he began shaping from gypsum or plaster of Paris a series of eighteen mathematical surfaces of the second order—ellipsoids, hyperboloids, and the like. A local publisher made copies for schools and colleges. The models thus went far and lasted long. Some twenty years later student Diesel's models found their way to Chicago's first world's fair, and as a "scholarly exhibit" were awarded the bronze medal at the great Columbian Exposition.

At the *Polytechnikum* Rudolf somehow found time to attend lectures on aesthetics and Goethe's poetry, and also to attend concerts. At this time he formed a lifelong admiration for the poetry and grandeur of the works of Richard Wagner. The shy student also reveled in Wagner's energy and mysticism.

In January, 1877, Rudolf Diesel, on entering his nineteenth year, now become an adult by government reckoning, received his German citizenship. For good measure the military authorities granted him a scholar's deferment; later, what was diagnosed as an aggravated asthmatic condition placed him in the "unfit for military service" category. There appears to be no record of his having suffered extensively from asthma; one may almost guess that the medical examiner may have leaned to the side of sparing an especially promising scholar.

That spring Rudolf's parents finally gave up their business in Paris and moved to Munich, bringing along their daughter Emma. Rudolf moved into their home at Schommerstrasse 3. For a short time Theodor found employment in a leather-goods store, but soon he gave that up to read and to attend lectures at the Polytechnic High School. Then "Papa T." became convinced that he possessed peculiar magnetic qualities and so could cure certain diseases. He presently practiced as a *"Heilmagnetiseur"*— magnetic health restorer—in Munich, using his home for his office. Rudolf did not offer his opinion, but one gathers that he rather wished that Papa T. had followed the whimsey trails of Grandfather Johann. After all, catching butterflies, feeding boxfuls of caterpillars, or holding harvest festivals to delight one's friends and honor the fruiting of one's lone apple tree are more socially acceptable than making believe one can cure the sick by way of "spiritual magnetics." Instead of objecting, however, the son continued his studies.

One of the most distinguished scholars at Polytechnic High School of Munich was Professor Carl von Linde, an authority on heat engines as well as refrigeration machines and a lecturer on textile machinery. There is good reason to believe that while listening attentively to a lecture by Linde, Rudolf Diesel conceived rather specifically the internal-combustion engine which would become the very special Diesel product.

The elderly teacher, who had developed according to his own principles an already successful refrigeration machine, told his listeners that steam engines used effectively but 6 to 10 per cent of the heat energy produced by their fuel. The thought of such tremendous waste so stirred the twenty-year-old Diesel that he made a notation on the margin of his student notebook, "Study the possibility of practical development of the isotherm."

There is evidence that this note was the nucleus of the principle of the efficient internal-combustion engine that was to change the world. But it was a long time before the idea was fully developed and incorporated in an actual working machine. However, if it is desirable to fix a definite time for the origin of the idea of the Diesel engine, that lecture on thermodynamics in the stately Munich hall of learning is perhaps the correct one.

Rudolf's notebook for July, 1878, indicates that he was greatly concerned with improving the heat-power ratio of engines then in use. He wrote about "Mechanical Heat Theory" and underlined, "Can one construct steam engines which carry out the complete cycle process without their being too complicated?" Two days later he jotted: "Would the efficiency of the steam engine not be improved if one would pump the water, not directly into the kettle, but into a tank, from which it would easily flow into the kettle, which, however, during the pumping is not connected with the kettle but with the atmosphere?"

The following school year, under "Notices about Heat

Theory" he wrote: "The Mechanical Heat Theory teaches that only a portion of the inherent heat of a body becomes available for outside work. One kilogram of hard coal combusted can convert water into steam having a heat value of 7,500 Cal. (1 calorie = 3.968 British Thermal Units), of which we can utilize only a minor portion of the energy for mechanical work. Does it not indicate that utilization of the steam or any intermediate body is unnecessary? The thought then follows to change this 7,500 Cal. directly, without transfer, into energy. But is that practically possible? That has to be found out!!'"

There was a great deal else to be found out, much of which related directly to Rudolf Diesel. For example, would his health, never robust, stand the strain of his energetic studies, plus his ventures in tutoring French and English, plus his daily tussles with piano lessons?

In 1879, as the naturalized citizen rounded twenty-one, he faced fair prospects of growing into a healthy man. His letters and personal records made occasional mention of headaches, toothaches, and earaches, as well as digestive upsets. But he ate sparingly, drank temperately if at all, and kept up a good attendance record.

Meanwhile, he gained good size, a rather lean six feet or perhaps a bit more—he later listed his ultimate height as six feet, two inches, but explained that he kept right on growing. Certainly, he managed to avoid the pasty pallor which was a characteristic of German and other European students of that era. One decisive factor here was his habitual walks. He walked in the sun, usually alone and frequently far. By gradual stages he had ceased to moon along, halting every few meters looking as if he listened to wonderful voices in the sky or the clouds. As he walked into manhood, Rudolf's stride grew longer and more rapid, his posture more erect, his mood more continuously thoughtful.

In the outreaching, fast-expanding field of mechanical engineering which he pursued so steadily, new findings appeared like bright, shooting stars. The most exciting entries were related to engines and power, still the crucial weakness of the new era of machines.

Half a century before, the French scientist Sadi Carnot had propounded the theory of a perfect engine cycle with the greatest possible thermal efficiency based on use of heat within the working body of the engine. Carnot insisted that the heat should be supplied at constant temperature.

Rudolf Diesel reasoned from this that the heat at the beginning of the cycle should be at a high temperature. The heated substance should then be expanded and the heat used without any loss to cylinder walls or atmosphere. The heat remaining after expansion should be extracted at the lowest possible temperature, which might be that of the atmosphere, then the air recompressed, without any loss, up to the combustion temperature of the fuel. Obviously, such a concept depended on mechanically accurate compression devices, which, practically speaking, were still lacking.

Another previously developed concept that interested Diesel was Beau de Rochas' outline for the four-stroke-cycle engine which had set forth the principle for the internal-combustion engine. The principle, per se, was quite simple: the downward suction stroke of the piston drew in the fuel and air. The second, the upward or compression, stroke compressed the mixture, which ignited approximately when the piston reached top dead center. The third, the downward power, stroke expanded the burned gases, while the fourth, or upward exhaust, stroke ridded the cylinder of the burned gases, thereby clearing the way for a new cycle. In this functional sequence Rudolf Diesel envisioned a virtually limitless source of power procurable from practically any burnable material—from whale oil or tallow to coal and what the amazing Americans were beginning to call "coal oil," or petroleum.

In his final year at Polytechnic, Rudolf Diesel studied even harder, spending more and more time with notebooks, in the laboratories, and in the school's small but efficient machine shop. To conserve his Sunday suit, he reported to classes and lecture halls in the blue work clothes which were the unofficial uniform of the Bavarian laboring man. Now that he was again living with his family, he studied even later and at times entertained his frequently ailing mother with improvised piano recitals. Like her, he preferred the German masters, including Bach and a more recent favorite, Brahms. But he also practiced the *Lieder* of a melody-suffused Austrian, Franz Schubert, whose music Germany was by then beginning to discover and enjoy.

Final examinations at Polytechnic were scheduled for July. In April, Rudolf found time to greet the spring and its glorious onrush of new life. In May he began preparing for the "grand" examinations. Although the school was still young, in its tenth year, the final examinations were already notorious for being hard and prolonged.

Rudolf's determination to lead his class remained unwavering. Then, as July came, the most promising of diploma candidates found himself ill; this time not with migraine headache or earache or indigestion, but with "typhus," then one of the most dreaded and fatal of all the cruel epidemics.

"Typhus" is the usual European designation for what Americans generally call typhoid fever. The 1964 Zermatt epidemic was a recent example of epidemic typhoid resulting from water bacteria. In Diesel's time Munich experienced a yearly outbreak. As Rudolf discovered, even the onset of typhoid can be painful. At the time there was no surely known preventive and no effective treatment other than bed care, prayer, and waiting. Certainly, Papa Theodor's magnetism was of no avail.

Survival usually depended on general health; the frail frequently died, and the strong were lucky to survive. Rudolf considered himself in the latter category. Great numbers of fellow Germans were not so fortunate. In Augsburg, Cousin Betty, wife of the hospitable mathematics professor, was among the casualties. One gathers that Rudolf, whom dear Frau Betty

Barnickel had joined her husband in designating as "that dear boy," barely pulled through.

By October, Rudolf had sufficiently recovered to seek out his first regular employment. He found work as an apprentice at the *Gebrüder Sulzer Maschinenfabrik* at Winterthur, Switzerland. After nearly three months, he returned to Munich for his final examinations.

Scheduling the examinations for January was actually a special "call," with professors assembled solely to question R. Diesel, who responded most ably. He not only passed the examination but achieved the highest grade average to have been scored by any student since the establishment of the school ten years before. It was a really memorable achievement. Rudolf's most famous mentor, Professor von Linde, was among those who were convinced that by all rights something should be done about such a record.

Herr Professor had influence in several quarters. He recommended honor student Diesel for an engineer's post in the Hirsch ice factory and refrigerator plant in Paris. But as it turned out, there was a vacancy in the Sulzer factory in Winterthur where the Linde refrigeration machines were being built.

Rudolf Diesel happily packed his shoulder valise, kissed his father, mother, and sister Emma good-bye and returned to Switzerland.

Young Man Makes Good

As MACHINE-BUILDING ESTABLISHMENTS WENT in those days, the Sulzer works of Winterthur was impressively large. It employed more than three-hundred men and built steam boilers, tanks, pumps, and many other machines made of cast iron and other metals. The list included refrigeration machines, as earlier noted, as well as valve gears used on steamships and paddle wheels for steamers. Sulzer *Maschinenfabrik* was an excellent place for a mechanical engineering student to grow into a mechanical engineer.

Rudolf Diesel returned to the plant resolved to be a real member of the work force. He moved into a factory-side rooming house where he joined a cluster of fifteen male engineers, shopkeepers, and office workers.

Discreetly, the graduate engineer donned his blue coveralls and reported for his first work assignment. As he expected, it turned out to be a tough one—filing out a giant screw key. But young Herr Diesel completed the assignment competently within a week. Then he worked a power lathe and a drill press before moving on to the screw-cutting shop.

In a very short time Rudolf found himself a fully initiated factory worker. He was never to lose his interest in or compassion for factory workers or his inclination to humor them. He definitely liked factory work. The machines were complex and demanded skill, but Diesel found himself liking the machines individually and collectively. To him they were creative things and challenged creative efforts.

And the machines had lessons to teach. As a student, Rudolf had clearly sensed this fact. He had learned much in addition to the contents of textbooks and laboratory manuals, more even than his crowded notebooks revealed. Earlier his parents had urged that he learn and earn as a mechanic, not as a scholar. He was now doing both, as he later recalled, with an inventor's mind but for a factory hand's wages. During his school years he had learned in general how to operate several metal lathes and other powered machines. At last he was applying in a practical manner what he had learned in one of the best-equipped machine shops then in existence.

There is good reason to believe that young Herr Diesel entered his first full-time factory employment with the ambition of inventing an engine suitable for small industries or for operating a single lathe or similar machine, since the smaller engines then in existence consumed too much expensive fuel for the work they accomplished. The steam engine was still too expensive and too wasteful for economical use as a small power unit. Steam power was, in truth, a monopoly builder. Only big industries could afford it.

The young man in blue cotton twill—*blaue Monteur* was the current phrase—foresaw a comparatively small but highly efficient engine for use in small shops, even home workshops. He pictured as prospective buyers dentists and watchmakers, woodworkers, and builders of bicycles and wheelchairs. He also anticipated the need for an engine to be used on sewing machines and washing machines and to power meat slicers and vegetable peelers for restaurant needs. At this time the electric motor was still in the future, and, indeed, only a few major cities had acquired electric power systems.

As he thought ahead, the young man worked well and continued to learn from the many and far-scattered machine shops of the great Sulzer plant.

He was especially interested in following the course of the individual parts as they moved through many hands and diverse operations in the process of assembling a particular machine. He observed and learned carefully the complexities of making and assembling parts for the refrigeration machines, for they were Professor von Linde's invention and, quite probably, his favorite. Professor von Linde had been Rudolf's instructor and most helpful friend. Rudolf was more than Linde's protégé; Linde's great inventions were to be Rudolf's professional and personal responsibilities.

In March, 1880, Rudolf was assigned to Paris to assist in the establishment of a new refrigeration plant. Baron Moritz von Hirsch, an international financier whose interests included railroads in the Orient and real estate speculation in many countries, had acquired French patent rights for the Linde machines and undertook to build a factory on the Quai de Grenelle, then burgeoning as a complex within the capital.

Baron Hirsch was interested in gifted young men but opposed overpaying them. Diesel's beginning salary was one hundred francs a month, conducive to a return to twelve-cent dinners. The following year, however, he was promoted to plant manager with the respectable salary of two hundred francs. By August this was doubled.

Hirsch's other interests soon took his attention from the Paris refrigeration plant, so he returned the sales license to Professor von Linde and sold the factory to Monsieur Fabry. As a result of this transfer, Rudolf Diesel took over the sales rights to Linde's refrigeration machines for France and, later, for Belgium. The equipment was highly rated. Several models of the "cold-making" machines were exhibited at a trade exposition in Versailles, where the machine received a gold medal award.

Diesel was once more a very busy young man. He had factory work to attend and refrigeration machines to sell and distribute to the most favorable markets. For good measure he supervised many installations and counseled on their proper operation.

During this time he was involved also in building an ammonia engine, the idea for which had first occurred to him while he was working at the Sulzer plant in Switzerland.

Diesel's first patentable inventions involved low temperatures, which he termed the lower shank of the mercury column. His first goal was the production of "crystal clear" ice suitable for bar and restaurant use. France, particularly Paris where there were many restaurants, was a potentially splendid market for table ice. As a noteworthy prelude, Diesel had developed a process for freezing small quantities of ice in bottles or jars. The technical requirements were formidable, but the industrious young man had no doubt that he would reach his goal.

"I will work like a dozen Negroes," he wrote his sister Emma in Munich, "but I tell you I will be successful." A short while later (October 11, 1881) Diesel asked his sister, then his closest confidante, to send him two hundred marks. Emma complied promptly, but with the money included some gratuitous advice. Sister-like, she suspected that he had fallen into the clutches of some unscrupulous woman and counseled that he mend his ways. Rudolf promptly answered that he needed the money to secure his first patents for the "ice-in-bottles" (*carafes frappés transparentes*) process and for making clear ice.

On September 24, the twenty-three-year-old inventor had received a precious certificate from the French patent office—his first patent. This patent for the "ice-in-bottles" process was followed a month later by one for the "crystal clear" ice process.

Rumor had it that there were private items not wholly crystal clear. It was reported by the customary indirection of extremely personal letters that young Rudolf had a mistress. The allegations were vague. She appears to have been an American divorcée whom he had first met in Munich. By odd happenstance she had followed him to Paris. She was said to have described Rudolf as a refined young man with an artistic temperament and a mind far above the generality of young men. When the female American appraiser left Paris, Rudolf wrote of his desire to go to the United States, which, so he said, held attraction for him "as the needle of a magnet to the North Pole."

But invention, for the time being, was his magnet, and Paris remained his place of work. He wrote of occasionally attending artists' parties in the Montmartre. But most of all he worked with a perceptive eye on the ways and wishes of the *bon vivant*.

Rural and urban Europeans were disposed to keep with the fine old tradition of sawing out storage ice in winter and watching it melt away the following summer. But France, especially Paris, showed appreciative interest in manufactured ice. German interest in refrigeration, particularly for foods and drugs, was rising. In addition, breweries were interested. R. Diesel was diligently developing these interests.

Professor von Linde, though still warmly interested in his protégé, was extremely cold toward the production of crystal or table ice. Diesel next wrote to the Sulzers, his former Swiss employers. They expressed interest in the young inventor's table- or clear-ice-making machine, but nothing came of it.

Diesel next got in touch with Heinrich Buz, then director of the *Maschinenfabrik* in Augsburg, who offered to build the machines and suggested that the inventor spend several weeks experimenting with them at the factory.

The pleasant and direct association of Buz and Diesel was later to prove important. Within a year some of the parts for the Diesel Clear-Ice Machines were being made by the Augsburg factory. Other parts were being produced in Munich at Rathgeber's Wagon Factory. Diesel managed to find time to supervise the first installations of the assembled machines at two testing points in France, a brewery at Châlons and the renowned Grillon Frères Châteauroux breweries.

The installations were successful, but to his great distress Diesel found he could not profit from his own ice-making machines. Even though he held patents which attested that his clear-ice machines were "new" and his own invention, French law stipulated that since Diesel was an employee of Herrn Professor Carl von Linde, the financial rewards must go directly to his employer.

The following year, evidently after painful thought, Diesel reached a no less painful decision. Since he could not change the law, and since he was quite certain he could invent other worthwhile machines and processes, he would resign from his employment, much as he enjoyed it. Therefore, with a rather unusual mixture of regret and resolve, he severed employee connection with Carl von Linde. Graciously the aged professor invited his protégé to stay on as a director of the refrigeration plant in Paris. The director's honorarium of 3,600 francs a year seemed ample for Rudolf's still modest needs. However, this, too, proved to be a temporary state of affairs.

Rudolf eventually took over the sale of Linde's ice-making and refrigeration machines as an independent merchant, thereby becoming, as the French say, an *ingénieur civil*. For this undertaking, too, there was a convincing reason. Her name was Martha.

Her Name Was Martha

As THE 1870's prepared to give way to the 1880's, Paris was regaining her warmth and verve. The great river town, which has seen a thousand tyrants fall and a thousand and one heroes rise, was once more becoming livable for her children and alluring to tourists.

Diesel had rented a modest apartment, "bachelor quarters," at 40 Rue de Rivoli—two rooms, pleasantly furnished, looking out on a red clay court and a tall elm tree, which the youthful second-floor tenant repeatedly sketched as an improvised letterhead.

Despite his preoccupation with inventions and commercial works, the young man with an obvious desire for the finer things of life was, at long last, liking and enjoying Paris. Although he continued to lodge alone, he was exhibiting a talent for making friends. No longer cramped for money, R. Diesel, by slow, sure stages, was overcoming his shyness. He invested one hundred francs in a deftly tailored blue wool suit and half as much again in a brightly ribboned felt hat with an American label—Stetson. The headgear may have symbolized his continued interest in Americans, including several among his Paris friends.

In his busy routine of work, Diesel continued to make many other friends. The list included factory workers, foremen, salesmen, mechanics, fellow engineers, and shopkeepers. Among the last was a prosperous Parisian storekeeper, Ernest Brandes.

In addition to his apparel shop, Monsieur Brandes had a city home, a charming wife, two young children, a governess, and a penchant for mid-week dinner parties. For the success of the dinners, an extra and unattached young man, particularly a good-looking and gifted young bachelor, was distinctly an asset. Diesel was happy to be included.

He did not dance, but he spoke and could understand French, German, and English quite ably. He was developing competence and charm as a conversationalist, and his voice was almost as pleasing as his smile. He sang passably well, and his piano playing qualified him as an impromptu drawing room *artiste*. During his Polytechnic School years Rudolf had learned to read music well, including the light concert pieces of a romantic young Norwegian named Edvard Grieg. And he knew Wagner.

During his first venture at Thursday night partying at the Brandes', Rudolf met the children's governess. In Paris, governesses were regarded with respect. Along with being a practical nurse, companion, and dutiful caretaker, the governess by Parisian interpretation was primarily a teacher, tutor, or supervisor of education. Her responsibilities were important, her educational requirements were extensive, and her social status was fairly secure. Socially, she was definitely acceptable as one of the family.

The Brandes' governess was German—Martha Flasche of Remscheid. Her family was respectable middle class; she was a notary's daughter. Largely self-educated, Martha Flasche, like Diesel's mother, was an excellent linguist and an aspiring cosmopolitan.

Fräulein Flasche spoke high-level German, even if not the enviably "pure" *Hannoveraner,* creditable English, and fluent French. No doubt R. Diesel found this refreshing. During his seven years in German schools, he had acquired a marked proficiency in his ancestral tongue. On his return to Paris, he had regained facility in French. By October, 1882, when he met Martha Flasche, he had shed practically every trace of what he sometimes termed the *Strudel Deutsch* accent—the American equivalent is "cement-mixer French." By now, in fact, he spoke French so well that he was being accepted as a native; literally, of course, he was.

Diesel confided that he was delighted and amazed by the poetic quality of Martha Flasche's German. There is no reason to believe that his delight and amazement were limited to the governess' linguistic talents. She was a very pretty young woman—German blonde, with deep blue eyes, flaxen hair, and a beautiful figure. Invariably her apparel was modest and tasteful. She was gay and witty. She sang quite passably in French and English as well as German; moreover, Rudolf enjoyed providing piano accompaniment for her singing.

Their common interest in music led directly to common interest in art and books. For a woman in the 1880's, Martha Flasche was remarkably well read. She had followed the new British novels and the great nineteenth-century poets of England, Germany, and France. She knew the works of the liberal French writers, including Rousseau and Victor Hugo, The two were, as Diesel observed, broadly and voraciously interested in art. As the year ended, the couple shared an experience which was beginning to transcend their common interests in books, paintings, and languages. They were falling in love.

The event, as many of Rudolf's friends saw it, was not especially advisable in terms of an ambitious young man's plans and prospects. None denied that Martha Flasche was pretty, witty, and delightfully feminine. But she was German.

Although a German citizen, Diesel was Paris born, and by now he was being accepted as a Frenchman. When the young man, who was beginning to describe himself as the contented iceman, occasionally spoke at trade gatherings or institutes, his French was fluent and convincing.

Although he spent freely on developing inventions, his financial prospects were bright. During his twenty-fifth year Diesel's earnings climbed abruptly to some 33,000 francs, somewhere near the present-day equivalent of that many dollars. He showed no indications of becoming "money spoiled." His living habits were not quite as conservative as they had been while he was earning 100 francs a month, yet even with an income averaging thirty times that amount, he continued to work capably and almost continuously, and to fall in love.

He noted that the latter phenomenon appeared to run in the family. In Munich, romance was also budding. After three years as a widower, Christoph Barnickel, Rudolf's ever gracious champion and counselor, began courting Rudolf's pretty young sister. It was a romance of kissing cousins which promptly flowered into marriage.

Herr Professor was in his late fifties. Emma was in her early twenties. But the newlyweds seemed blissfully happy. Emma, for the first time in her life, was secure—no longer the pretty young shoe cleaner or the seeker of menial jobs. And Rudolf's favorite cousin by marriage was now his welcome brother-in-law.

Spring found the ingenious young iceman back in Paris and wooing. With Martha at his side he again walked the suburbs of Vincennes, the somber halls of the Louvre, and the dusty aisles of the *Conservatoire des Arts et Métiers.*

Rudolf had suggested a May Day marriage. Martha, who appeared to favor a longer engagement, returned to Germany for a visit with her own family and her fiancé's.

Diesel set forth to spend the summer at work in the provinces. A principal job was installing an ice machine for the Grillon Frères brewery at Châteauroux, Indre. He urged his work crew to keep on the job over the week end, which happened to be the time of Corpus Christi. However, the workmen declined to work on that holy day.

On May 16 he wrote to Martha, who was visiting Theodor and Elise Diesel in Munich: "When I close my eyes and think of the last three days of Pentecost, I feel as if my body becomes lighter and lighter, lifts itself up from the earth, not any more physical, and floats in unlimited ether through which rays of the sun spray a rain of flames, in which the chirping of birds, the hum of mosquitoes, the sound of wind, create an indefinite but intoxicating music. Within this, a sweet, sweet angel voice, saying, 'Rudolf, I love you.' "

He spoke of his poetic vein, which was now a weaker one in the battle against his *Eismaschinenader* (ice-machine vein). Then he hoped that Beelzebub would appear and might bring thunder and lightning to pulverize the miserable nest (Châteauroux) so that he could come and embrace his Martha, fall around her neck, and listen to her voice and tell her of his unlimited love and kiss her, and then again return to the thirty masons and carpenters and tinsmiths . . . .

On her journey to Germany, Martha had passed through Châlons, where Rudolf was working on another brewery installation. Shortly after this, on June 10, 1883, he wrote to her. "It is midnight! All alone, silent, from time to time gazing down excitedly on the empty *perron* [platform]—. Then the machine roars into the station. The train stops. All of the travelers are deep in sleep; one lone window opens; I speed towards it, to open the door, and into my arms sinks silently an angel appearance, kissing me enthusiastically on the lips, while I lock her tight into my arms.—I love you, my girl! I love you, I whisper into her ears! Good-bye [*Leb'wohl*], greetings to all. Another kiss—and it goes on again. Before I come to my senses, the train has again disappeared into the night; I watch it eagerly. Was it a dream? No, a warm tear on my hand brings me back to reality; be a man, I say to myself, and speedily I wander home through the dark streets of the city which appear to me even darker and more lonesome and lost than before."

After a three-week period of acute loneliness and painful separation from his loved one, he wrote to her at Berchtesgaden: "Yes, be my little star and lighten the darkness, the deep darkness of my heart; my heart is like a deep, dark, unlighted lake, surrounded by black fir and rocks, which is so black at night that one cannot see it. But if such a little star comes with its golden sheen, then he mirrors it a thousandfold in his waves and it is not dark any more, but shimmering and alive from the light of its star. To the Königsee you will probably come soon, to the unlighted, unlighted Königsee. But don't fall in; there it is deep, and when one is in it, one will not be saved. If you drown in it, I will immediately come to you and search for you in the deep and remain with you."

Later he wrote: "Only two or three more letters, and I will then only write, I come, I fly to you, into your arms, out of which I will then never go for the rest of my life."

They were married before members of the family on November 24, 1883, in Munich and moved to Diesel's Paris apartment, 40 Rue de Rivoli. Acceptance of Germans in Paris was again chilling markedly. Not wholly by coincidence, Diesel's income was shrinking and so was his circle of friends. His friends, among whom were some Germans and Americans, also included French acquaintances who disapproved of his marrying a German girl. Although by no means wholly ostracized, the bride and groom had fewer friends. But they had each other, and about a year later they had a son Rudolf.

Early in 1884, Louis Phillipe Cohen, an able merchant from Hannover, invited the groom to join him as a partner in forming a distributing company to sell refrigeration machinery. Diesel accepted, but the venture did not succeed. His income continued to plummet. For this and other reasons, the glamour of Paris seemed perceptibly on the wane.

One reason was the increasingly hostile attitude toward Germans. Among the causes of this was the attitude of General Boulanger, who was proving himself more effective as a jingoist and arouser of nationalism than he had been as a leader on the battlefields. The heralded rise of the Third Republic of France could not completely cure her Prussian war wounds. "Anti-Germanism" was reaching avalanche proportions.

On May 22, 1885, at the advanced age of eighty-two, the great Victor Hugo died. There followed an interval of national mourning throughout France and the entire literary world. His writings had stirred men and women everywhere. The body of the poet-novelist lay in state under the Arc de Triomphe. When Rudolf and Martha attended the services held at the great arch, both were dismayed by the cheap carnival atmosphere and the lack of dignity. Hawkers sold pictures of the eminent man of letters engraved on coins, brooches, pins, rings, and other jewelry items. A small group sang a song, an alleged favorite of the dead author, and sold copies of it. Candy sellers, cake vendors, sweet-drink vendors, and other hawkers shouted at the tops of their voices, advertising their cheap wares. How immaculate in truth was the immaculate soul of France?

Martha Diesel conceded that her low spirits may have been accentuated by the fact that she was in her second pregnancy. Their second child, a daughter whom they named Hedy, was born on October 15, 1885.

For the young father that date was doubly memorable. A factory which he frequently visited, one which had been helping with his proposed ammonia engine, suffered an odd sort of accident. A container of ammonia exploded. There were no fatalities, but workmen were disabled by the release of the benumbing gas.

Ever observant, Rudolf Diesel seized on the idea of bottling the powerful gas as a military weapon. Instead of bombs and bullets a combat force could throw or otherwise release small vials of ammonia—to incapacitate temporarily an enemy without killing or permanently maiming him. As a good German, Diesel took the proposal to Graf Münster, secretary of the German consulate in Paris. Count Münster laughingly rejected the idea. Thickheadedness never was an exclusive attribute of any particular nation, Diesel mused.

The new father kept finding additional evidences of such stupidity. He continued to meet prospective customers who would not buy his wares simply because the Linde and related ice-making machinery were supposedly of German origin. Even the breweries, then the most logical customers for refrigeration equipment, were demonstrating anti-German sentiment. Pointing out that most of the machinery was still being manufactured in Switzerland did not greatly help with the impasse. In the rural and small-town areas, artificial ice-making lagged because the previous winter had been unusually cold and the "natural" ice harvest had been heavy.

When his wonderful new daughter was duly "situated," Diesel resolved to set forth on a selling trip to a land where freezing winters and determinedly energetic farmers did not conspire to fill the ice houses brimful. He traveled to French North Africa. In Algiers he found tremendous need and demand for ice machines, but there was no money with which to buy them.

Back in Paris the Boulanger-voiced anti-German witch hunt was more pervasive and no better smelling than the prevailing sulfurous coal smoke. Diesel admitted his discouragement. He was eager to return to his proposed ammonia-using engine, an invention by now six months neglected.

He still believed that using ammonia vapor in place of steam should and could significantly raise the work efficiency of the steam engine. As his thirtieth birthday neared, the young man with a purpose, as well as a growing family, settled down to intensive work on the ammonia engine. As practice had already proved, when Rudolf Diesel said "intensive work," he meant it. He rented a vacant machine shed near his Paris home and employed four men to help bring into workable form, the already proposed gas engine. He had resolved to complete the task if it took all summer. It did take all summer, and much longer.

The young inventor rechecked his design and settled to a marathon of work. His helpers labored the prevailing ten-hour day. His own workday was nearer twenty hours. He had all his meals brought to him and slept only a very few hours each night. As the weeks wore on, fatigue began to tell. Time and time again, the inventor was obliged to sit or catnap while the work went on. His urge to complete the ammonia engine grew to be a virtual obsession.

Costs ran heavy. In course of the summer Diesel was stricken by severe and chronic headaches with migraine symptoms. At earlier periods of his life, most memorably during his family's desperate exodus to London, he had suffered excruciating headaches. But the current ones were the most punishing he had ever known. One cryptic doctor suggested his headaches would subside when his invention worries were alleviated. That was in the nature of a distant assurance, for his invention worries were actually growing.

Headaches are a complex phenomenon. In Diesel's case there is reason to infer that some part of the cause was emotional. From all appearances his marriage was still happy. Martha had accepted understandingly her role as an "inventor's widow." But there was evidence that both the Diesels were increasingly upset by the anti-German mania which continued to harass Paris and most of France. As the months followed, Rudolf saw his hopes for prosperity fading into the lingering red dust of Paris.

Aside from the man-racking work hours and the accompanying strain on muscles, eyes, and mind, Diesel was plagued during his work with ammonia by its chemical and physical qualities. The biting, highly poisonous gas would repeatedly escape, even through invisibly tiny holes in engine parts or feed lines, to cause great discomfort, including eye and throat burn and intense nausea, to all the workers.

Fumes or not, and decimating headaches bedamned, Diesel kept at work. By summer's end he had developed a working engine.

Earlier that year at a meeting of the *Société des Ingénieurs Civils,* Diesel had heard that the Marquis de Montgrand had invented an apparatus which functioned according to thermodynamic principle. On his copy of the Montgrand report, Diesel underlined these words heavily: ". . . that compressing gas develops heat during the reduction of its volume. If one increased the original volume, heat is absorbed, being removed from the neighboring bodies." This might very well have brought back the memory of the small fire pump at the Augsburg school and suggested anew the possibilities of an engine to be driven by heat engendered by the process of compressing air.

Later that year Diesel splurged on an outing in Hindelang in the Allgäu, one of the more appealing resorts for the tired and the bored. Diesel was certainly not bored; he was painfully and benumbingly tired. Hindelang was a wonder spot made for rest of body and revival of spirits.

The hideaway is only one hundred miles or so from Munich. Even then the fast train to Sonthofen brought one near the *Kurort,* which is situated about 2,500 feet above sea level almost at the end of Ostrachtal. There are nearby a fast-flowing river, sunny mountain slopes, and narrow valleys which steep mountainsides shelter from raw winds. The resort region is traditionally sunny, because the piling fogs are shielded away. And the area is a hiker's paradise. Series of trails lead deeper into the Ostrach Valley to yield magnificent views of the snow-capped Alpine peaks which surround them. A twisting road leads up the steep mountain grade into Austria.

In this secluded spot the tired inventor found solace for his toils and worries. With his family and at times alone, he walked in the healing sunshine. In this happy exercise he conceived, perhaps "dreamed up" would be a better phrase, an engine to be powered by the sun's energy. Apparently he drew no diagrams of this grand idea. Certainly he committed little data to paper. But he envisioned a power-developing device using sunlight whereby air would be heated in a large iron drum to a high temperature and the expanding heated air would drive the engine's piston; when cooled by water, the air would contract and the piston would go back to first position.

The solar engine, of course, is a recurring daydream of the thermo-physicists. Diesel's Hindelang rhapsody, however, seemed especially revealing of the creative intelligence of one student and philosopher of thermal phenomena. Practically defined, the field of thermodynamics was as new as it is vast. History tells that most memorable inventors stay close to a single *new* subject or field of interest. At the present interval of history the space engine is the pre-eminently new subject in its ageless and infinite field. Were he living today, Diesel quite probably would be an inventor of space engines. However, even in the late 1800's Diesel was creating a power unit many decades ahead of its time.

On the practical and immediate side, the need for heat engines included such pragmatic challenges as developing a "cool box" for household use. Its refrigeration would be made possible by a small gas flame, employing the now familiar thermal principle of the home refrigerator.

Time and time again Diesel heard his proposed inventions damned with faint praise—"interesting but impractical." At least in this discouraging and monotonous appraisal R. Diesel was not a lone target. The entire era was shaping up as one in which dedicated inventors were being pooh-poohed or more abusively reviled by practically everybody.

Meanwhile, Paris was again grooming herself to be hostess for another world's fair, already proclaimed the greatest in history. The opening was scheduled for 1889; the exact date waited the completion of construction. The most impressive of buildings was the Eiffel Tower. For two years Rudolf Diesel was among the hundreds of thousands of persons who looked on, at least occasionally, while the intricately laced iron and steel work reached higher and higher into the Parisian sky, presently climbing to the then unprecedented height of 300 meters (984 feet). Many Parisians had labeled the structure a monstrosity and forecast that it would collapse long before it could be completed. But Gustave Eiffel, who had previously built many "impossible" bridges, declined to heed the prophets of gloom. The master bridge builder kept on supervising his beloved and impossible tower, which he saw completed to the marked benefit of the greatest of Paris fairs as well as to his own deep satisfaction.

Fair opening marked several important events in Rudolf Diesel's life. For one notable item, his second son and third child, Eugen, was born in Paris, May 3, 1889. For another, he was authorized to exhibit the Linde ice machines at the great fair. Earlier he had planned to exhibit his ammonia engine, which was beginning to be operable. But for unknown reasons, he changed his mind and did not show his newest major invention. One can guess a reason for the change of plans. Although the efficiency of the "gas engine" was already well above that of the steam engine, the operation remained rather difficult and somewhat hazardous. Should the ammonia escape, as was still readily possible, injury could result to the operator or onlookers.

Diesel spent much time viewing the exhibitions of steam engines and of the more up-to-date internal-combustion engines fueled with illuminating gas. Indeed, Paris' Universal Exposition of 1889 was a landmark for engine making.

The first motorcar ever to be shown publicly, a three-wheeled Benz Patent-Motorwagen, stood boldly in a separate pavilion. Several Daimler-built four-stroke-cycle internal-combustion engines, using refined gasoline, ran continually and with a formidable uproar. A tramcar, powered by a Daimler engine, stood silent on short rails outside the engine pavilion. All the eye-opening exhibits contributed to the tremendous success of the fair, which was attended by more than twenty-eight million persons. Fireworks as well as fires confined in cylinders were among memorable entries.

In connection with the world's fair, an international congress for skilled mechanics was held in September. Rudolf Diesel was the only German who was asked to read a paper to the large assembly. He discoursed brilliantly on the *"Revue Technique de l'Exposition Universelle."* He spoke French so ably that few in his very large audience even suspected that the speaker was not a Frenchman.

Yet, when the great fair and all its festivities were ended, the Parisian prejudice against Germans and German-made goods persisted. Although the Linde machines were being manufactured by the Swiss Sulzer Brothers in Winterthur, they somehow remained listed and damned as German products. Even when a French company began to manufacture them, nationalistic opposition and sales resistance did not materially abate.

During November, 1889, Diesel went to Munich for another firsthand meeting with his old friend and mentor, Professor Carl von Linde. The protégé stated his willingness, indeed his eagerness, to give up Paris and return to Germany. Herr Professor smiled approval and maneuvered successfully to secure his still-favorite student a suitable position. Shortly thereafter Diesel found himself with a franchise to distribute and sell Linde machines in northern and eastern Germany, with a guaranteed base salary of thirty thousand francs a year.

Once more Rudolf Diesel returned to Paris to make ready to leave it. This time he was in good company. He had a wife who was still very pretty and a gay partner. They had three children who shared the physical attributes of an exceptionally pretty mother and an exceptionally handsome father. They had assurance of a bountiful livelihood and the likelihood of a happy life together. The husband had prospects of great success in his chosen work. Temporarily he enjoyed surcease from his ruinous headaches.

The Serious Inventor

BEGINNING ON OR ABOUT THE TIME of his thirtieth birthday, Rudolf Diesel decided to list himself as a "thermal engineer," instead of a mechanical engineer. More and more frequently he reminded his customers and miscellaneous listeners that cold is merely a degree of classification of heat, and refrigeration is but one practical use of the "thermo-engine."

He reiterated that ice-making was but one demonstration of the useful developments possible through mechanical compression of gases. The compression of air under high pressure could also produce heat, or usable power. The thesis was, as the speaker frequently pointed out, in essence a simple exercise in mathematics. When listeners sometimes observed that they did not find the young thermal engineer's concept of simple mathematics in any way similar to their own, Herr Diesel would smile engagingly and begin murmuring equations, accompanying them with slight but expressive hand gestures.

The inventor was still in his twenties when he received his first honorary citation. It was much more than the present equivalent of an honorary doctorate. By long tradition German schools bestow their honorary degrees, particularly in the sciences, with great care, frequently granting no more than one during a given year.

Diesel was aware of this. But his acceptance speech was distinctly gay and very far from being technical or scientific. "I am an iceman," he began. "One of the kind of icemen who attempt to produce the utmost coolness among men—in the form of ice, ice water, cool air, and all such as that. . . . I am settled in Paris, endeavoring to cool the spirit of revenge in the hereditary enemy of Germans. . . . I should like to do away completely with the heat of wars . . . ." The acceptance speech apparently was largely if not entirely impromptu. Certainly its sincerity was as apparent as its gentle wit.

In February, 1890, Diesel, then nearing his third-of-a-century mark, first set foot in Berlin, the "new" capital of the "new" Germany. The inventor-salesman was not especially glad to be a Berliner—even on a temporary basis. His new assignment as a tender of Linde concessions for northern and eastern Germany virtually required that he make headquarters in the Prussian capital; with reasonably good grace he capitulated.

As Diesel gained acquaintance with Berlin, he found himself liking certain features of it. The metropolis had restless vitality intermingled with a rather surprising and easy-going joviality. Other unexpected qualities, such as conscientious correctness and naïve sentimentality, added to the somewhat baffling compounding of the Berliner.

The grandiose *Kaiserstadt* was startlingly impressive. Wide, straight boulevards cut through the city with military precision. The streets were un-Frenchly clean, almost as clean as a good *Hausfrau's* doorstep. Fuming locomotives pulled trains, running on close schedules, on the *Ringbahn* around the city. From half-a-dozen railway depots sparkling new railroad trains puffed and rumbled forth to all parts of Germany. On many streets the horse-drawn streetcars ran on iron rails. Some of the more elegant homes had electric lights over their entrances. Already electric cables had been laid under the streets by the large Siemens concern. But in any city electricity was still sufficiently uncommon to attract attention. Gas remained the standard illumination fuel; as yet electricity bore a connotation of wizardry. Berlin was being electrified.

The Diesels took an apartment on the rather fashionable

Kurfürstendamm. The house, Number 113, had just been completed and was located almost at the end of the wide boulevard. From their windows the newcomers could look out on far-flung brown meadows and leafless trees.

The Berlin of 1890 was conspicuously a city for industrialists and businessmen, for the nobility and the military, and, most of all, for the Prussians. Many noblemen had married the willing daughters of the untitled wealthy. In great part the grooms were officers in the Kaiser's army. Neighbors of the Diesels included the officer families of von Kleist, von Kotze, von Motz, von Luckwald, and von Schack. Other neighbors and, with reservations, friends of the Diesels included engineers, professors, and industrialists.

The abrupt rise in the Diesels' standard of living was astonishing. In Paris, Rudolf had taken his bride to a very modest two-room bachelor's lodging and there stayed even after the birth of their daughter and two sons; and, at least during the first two years of their marriage, the groom's income actually exceeded his assured earnings in Berlin. It is true, of course, that his subsequent earnings fell to only a few thousand francs a year. His savings had apparently melted away. In March, 1890, when Diesel moved his wife, their first son, little Rudolf, their toddler daughter, Hedy, and Baby Eugen to Berlin and leased the apartment on Kurfürstendamm, he was nearly penniless; however, he committed almost half of his assured income to housing alone.

Even when elegantly housed, Diesel still did not like Berlin. After his many years in the gay French capital, he felt at first a foreigner in the sophisticated new capital of Germany. It was preponderantly a city in which citizens were segregated on a basis of vocation or trade. The military had taken over social leadership. As the young man from Paris saw it, Berlin factory workers were slaves of the machines they tended. In the main they were efficient workers; yet they were disturbed and ill-contented, and they talked of strange-sounding credos, including what was labeled Marxian philosophy.

Although he was a man of many sympathies, Diesel had no special sympathy for this particular vehement professor, economist, and writer. Karl Marx, Diesel noted, had based most of his views and findings on industrial England, not Germany. Even so, during the lifetime of Karl Marx, the Marxian onslaught against capital enterprise was widely noted and greatly admired by working-class Germans and, no doubt, some few others. At the time of Diesel's arrival in Berlin, Marx had been dead for seven years, but his so-called philosophy was very much alive. The ardent socialist's near contemporary Prince Otto Leopold von Bismarck-Schönhausen was quite impressively alive even at seventy-five. As the newcomer noted, in many factories and shops one heard more mention of Marx than of Bismarck.

Diesel was personally appalled by social contacts with the dominant caste, some part of which seemed purposefully dedicated to inciting and waging wars. At the time, Diesel believed with many Germans that a competent military force could serve to avert war and offer needed defense against encircling enemies, potential or real. Some of his beliefs changed later, but not his appraisal of Karl Marx. As a descendant of manual tradesmen, Diesel was disturbed by the Marxian image of a proletariat inevitably segregated. At this time Diesel was in a position to select his own social contacts. In doing so, he was not disposed to demonstrate acceptance of any particular class or "caste."

Meanwhile Diesel had accepted the triple role of salesman, earner, and inventor. For a surprising and most welcome change from his Parisian status, Diesel was being accepted in Berlin. He was recognized by professors and engineers as a young man of consequence in what Berliners were calling thermodynamics. In their curt, stiff-spined way the young army officer neighbors were cordial enough. Industrial leaders of the Prussian metropolis were similarly cordial. Almost immediately Diesel found himself elected a director of an incorporated association of cold storage and sales agency operators. This, in any of Diesel's three languages, was a long step ahead.

There was solid though subtle evidence that Martha Flasche Diesel was largely responsible for the family's improved social standing. The engaging young wife was finding in Berlin many elements to her liking. She had social potentials not to be longer repressed. She liked people and she liked money, and was determined to steer her husband toward more of both.

Rudolf was not unco-operative or deliberately opposed. Fortunately for his own readily changeable moods, he was not strictly confined to Berlin. He traveled widely to sell Linde refrigeration machines; he stopped frequently to supervise their installation or to repair breakdowns.

On a business trip to Frankfurt am Main, Diesel ran into his school friend, Oskar von Miller, who had just supervised building a 25,000-volt electric line from Lauffen on the Neckar to the Electro-technic Exposition at Frankfurt for the purpose of powering a 180-horsepower electric motor. At the time this was a remarkable, almost inconceivable feat. Diesel's immediate reaction was a renewed burst of interest in engines for creating power—in contrast to motors, which, of course, are for transforming energy already provided.

Diesel liked Miller. They had been friendly rivals in technical school. The thermal engineer felt an urge—frankly competitive —to get back to his ammonia engine.

The gas-fueled internal-combustion engines of Otto and Langen and of Daimler were finding users. Diesel was quite certain that his projected ammonia vapor engine had merit, at least in potentially efficient use of fuel. He was already convinced that fuel efficiency would prove itself the ultimate factor in the success or failure of any or all engine designs. In terms of fuel consumption, steam engines could never be efficient. There was no immediate prospect that either the electric motor or the illuminating-gas engine could be made inexpensive to operate.

During his years of experimenting with refrigeration machinery, Diesel had demonstrated to his own satisfaction that as an engine fuel, ammonia vapor was vastly superior to many other elements. In Berlin he proceeded to set up a laboratory workshop for resuming his study of ammonia vapor and other phenomena related to engine operation.

The studies led toward another trail of discovery. As an engine fuel, ammonia is troublesome and toxic. The desire to avoid its peril and nuisance unquestionably had a part in motivating Diesel's determination to prove that air could be compressed and with relative safety. Heat implies an energy flow from one body to another by reason of a temperature difference. In a confined space the energy flow could be tremendously enhanced by imposing fuel particles which the process of air compression could cause to ignite, thereby increasing the heat or power.

The idea was sound. After many months spent developing and supporting the "correct theory," Diesel applied for and received a patent (No. 67,207) for a work process for developing a "combustion power engine"—as named in multi-compound German, *Arbeitsverfahren und Ausführungsart für Verbrennungskraftmaschinen*. At first look the Imperial patent office vetoed the process as "not original." Diesel appealed for reconsideration. After lengthy re-examination, the *Kaiserliches Patentamt* presently issued (on February 28, 1892), what was called a development patent on the process.

There were important correlations. In 1887, Otto Köhler had written a book entitled *The Theory of Gas Engines*. In 1891, Emil Capitaine of the Swiderski machine works in Leipzig had worked on an engine which showed some similarity to Diesel's latest conception. Apparently, as of 1892, Diesel was not familiar with the work of either man. However, the following year, when he got in touch with Krupp, he specifically mentioned Köhler's patents. Subsequent investigation by Krupp engineers found no patent infringements.

Rudolf Diesel then tried to interest one or more important and resourceful factories in underwriting the construction of an engine according to his patented specifications. In March, 1892, he wrote to the *Maschinenfabrik* Augsburg, where another school friend, Lucian Vogel, now the son-in-law of the plant director, Heinrich Buz, was working. But Herr Josef Krumper, the firm's chief engineer, was a steam engineer, and he advised against the proposed air pressure or "blowgun" engine. Director Buz therefore turned Diesel down.

The self-labeled iceman was still determined to have the Augsburg factory undertake and underwrite his work. He had confidence in its personnel. He wanted to go home, and, more than any other place anywhere, he thought of Augsburg as home. The inventor pondered how best to influence the company to reconsider its decision. Time was of the essence. Patent protection covered only fifteen years. Diesel was well aware that he had in hand a long-term job. Completing a satisfactorily operating engine of such a revolutionary design would surely consume much time; introducing it effectively could take even longer. Unquestionably the sponsoring factory would be obliged to risk a substantial amount of money. Time would be required to recover the investment.

It was Martha Diesel who thought of the solution to the problem—to expand the idea of the proposed heated-air engine and to publish the concept. Diesel set to work at once on the project.

But before *Theory and Construction of a Rational Heat Engine to Replace the Steam Engines and the Currently Known Combustion Engines* appeared in January, 1893, the *Maschinenfabrik* Augsburg had decided to construct Diesel's proposed engine. It was now fourteen years since the young student had made the memorable notation in his college notebook at Munich.

The publication of the Diesel booklet aroused great interest and considerable controversy. In most scientific circles, the theories were severly criticized. Even so, several authorities of stature, including Professors von Linde and Schröter, Dr. Zeuner, and Eugen Langen, now director of the important *Gasmotorenfabrik* Deutz, were quite favorably impressed with the specifications of the proposed engine. With almost naïve candor and schoolboyish sincerity, Diesel admitted in his pamphlet that "complete usability" of the gases as engine fuel was only theoretically correct or possible. From the standpoint of practical application, no engine can be completely efficient. He also stressed that for a practical engine, maximum temperature is not the most important feature; maximum compression is.

The main features claimed for the engine patented by Rudolf Diesel were the heating of pure air in a working cylinder (via mechanical compression by means of the piston without any heat loss to the cylinder wall) to a temperature substantially above that of ignition temperature of the particular fuel used. The concept and design also provided for the introduction of finely atomized fuel into the compressed air which would develop heat by pushing down the confining piston through what Diesel termed "isothermal expansion."

The engine fuel would burn best in gaseous form. But in deference to practicality, Diesel had also pondered the possibilities of using non-gaseous fuels. To test their use he suggested the introduction into the working cylinder itself of minute quantities of fuel sufficient to produce only one stroke of the piston. This, in turn, would require the discharge of the burned gas and the beginning of another "routine of compression." The result was economy in avoiding the complex and wasteful work of "generating" the fuel gas. The engine which Diesel visualized would effect air compression and fuel ignition in one process.

If the air is sufficiently compressed, its temperature rises above the ignition point of a given fuel, causing the latter to ignite spontaneously. The ignition of liquid or gaseous fuels then in use was attainable in a "warmed-up engine," made possible by in-cylinder pressures somewhere between ten and fifteen times the atmospheric pressure at sea level. Diesel's own experimentation had proved conclusively that the highest combustion efficiency can be attained at thirty to forty atmospheres. The work challenge was that of compressing the air to its highest attainable heat yield, then precisely injecting the fuel.

Another important requirement was the measured admission of finely atomized fuel into the highly compressed air. Still another was to permit the burning of the admitted fuel to take place within the cylinder itself and to discharge the burned gases before compressing the fresh air.

Experiments were being made in other quarters at that time to utilize heavy oils as fuels. But many peculiar difficulties had arisen, including various and special problems of tar and coal deposits and chemical and physical properties of the oils in use. These, combined with the losses in the heating processes, made all of the many attempted solutions unworkable.

Later developments proved that the intricate problem of injecting and atomizing fuel was the most formidable stumbling block for the "Diesel process." This was a combined procedure for compressing air to establish a temperature sufficient to bring about auto-ignition plus an interrelated sequence of actions and reactions. If the statement sounded complex, the development would be even more so.

Diesel was too well aware of the fact that between the theoretically proposed engine as described in his booklet and the eventually practical machine was a very wide gap. The entire area of pressure measurement and the reaction of bodies under stress was a limbo of doubt and speculation. He foresaw that the theoretical efficiency ratios of the thermal pressure engine—up to 70 to 80 per cent—could not be achieved in practice. He also held the conviction that the practical attainment of half the theoretical ratio of efficiency would be a decisive victory in man's quest for mechanical power.

The proposed feat of compressing air to the ratio of 250 to 300 atmospheres, say 3,670 to 4,400 pounds a square inch, was far beyond any compression ratio previously attained. Actually, the ratio which Diesel recommended was at least six times the maximum yet arrived at by any internal-compression engine then known. One "atmosphere" represents pressure of about 14.7 pounds a square inch. As of 1890, about 550 pounds a square inch, or about 31 atmospheres, was the highest that had been measured.

The choice of fuel for injecting into the pressure chamber was a hardly less appalling problem. Diesel had taken it for granted that raw oils would be the most suitable fuels. He recognized that heat generation would require cooling of the engine, for which he favored the use of water.

He was certain that such an engine would be versatile and capable of many different uses. It could be comparatively simple to operate. It could be much lighter and less bulky than other types of engines then in use. Certainly the proposed engine would not require a firebox, boiler, smokestack, or other comparable encumbrances. It could power individual railroad cars. Aboard ships it could make available for cargo most of the space required for steam boilers, coal bins, and other ponderous paraphernalia of steam engines.

Already, and quite sincerely, Rudolf Diesel believed in the engine he was developing. He believed that such a prime mover could make what was being called the "New Industrialism" a success, and this without the enslavement of its workers. The inventor believed and argued with great conviction that a really effective machine for generating power was the best attainable defense against the bodily and spiritual enslavement of mankind.

Even so, while thinking and speaking as an engineer and mathematician as well as a craftsman experienced in installing, repairing, and tinkering with relatively complex machinery, Diesel sensed that many details were still to be worked out.

With laborious thoroughness he revised and expanded his graphic descriptions of what would presently be the Diesel engine. Fortunately for Diesel's purposes, the Imperial German Patent Office had permitted "protective assignments," not only of plans and ideas for useful inventions, but of revised plans and improved ideas for useful inventions. During November, 1893, the persistent young man was granted a second patent (No. 82,168) for an invention much closer to that which emerged as the Diesel engine. The second patent included a scholarly amassing of diagrams, statistical curves indicating fuel use, and designs for injecting fuel and compressed air into the cylinder.

Publication of the patent won worthy advocates for the engine design. These included respected professors and theorists and, of greater importance, practical engineers and plant directors of the *Maschinenfabrik* at Augsburg, Krupp of Essen, the Sulzers of Winterthur, and many others.

"𝔚𝔲𝔫𝔡𝔢𝔯𝔟𝔞𝔯"

THE SON OF THE ECCENTRIC LEATHER MAN had gone far in a comparatively short time. In several ways his progress was typical of that of his contemporaries or near contemporaries in the field of invention. Others of Germany's pioneer inventors in the engineering or automotive field, including Nikolaus August Otto, Gottlieb Daimler, Karl Benz, Wilhelm Maybach, and Robert Bosch, also came from middle-class working families. Their fathers were skilled tradesmen or merchants, including a small farmer, a locksmith, a train engineer, and a leather merchant.

The education of these inventors was progressively more advanced. Otto had merely a grade school education; Daimler somewhat better schooling; and Diesel a still higher education, which enabled him to study extensively this new science of thermodynamics. He recognized and understood the theoretical problems and tried to apply them practically.

To an already memorable degree Diesel was succeeding in the practical application. But he was paying the price, including, as he himself admitted, a regrettable division of affection. As Rudolf stated with his sometimes blunt-edged humor, he now had two wives—his lovely white bride, Martha, and his demanding black mistress, the laboriously materializing engine, called *"Schwarze Geliebte"* by their friend, Lucie von Motz. And to complicate matters, he had torturous headaches and gluttonous customers.

Repeatedly during 1886 the busy traveler had complained of the heavy drinking and piggish eating of some of his more profitable customers. He wrote his wife how he deplored evenings he was obliged to spend in fancy restaurants and the drinking which followed. In most unflattering terms, he described the director of the brewery at Nancy and his cronies: "fat pigs" (fette Schweine). While they staggered home after the drinking bouts, Diesel walked with a steady gait to the Hotel d'Angleterre. He still was not a drinking man, and he would not be made into one.

One suspects that his estimate of the fette Schweine of the French breweries may have colored even his appraisal of the scenery. Of the Rhone he wrote, "Close to the shore of the Rhone our train goes along, the river rolls golden waves in the glow of the setting sun. On the opposite shore mountains and castles; it is very beautiful, but does not have the poetry and appeal of our Rhine."

But there were still those damnable headaches. In September, 1887, the young inventor wrote: "I have a headache—to drive me mad, really terrible [zum rasend werden, wirklich ganz fürchterlich]." Yet the constant thought of his invention drove him on. "I speed from city to city so that I can hardly think of my home; I am busy all of the time with my engine and my future; day and night this does not leave me alone."

On March 18, 1888, Diesel confided: "I am now thirty years old; I feel like a very old person and am sorry by and large that I have accomplished and worked so little. When I consider what is behind me, it does not mean much [autant que rien]. On the other hand, I feel that my energy is growing, and so all hope is not lost . . . . I work like a horse, and if I am not impressed too much with my abilities, I believe that such a tremendous amount of work—I really try very hard—eventually has to bring results and that my striving [Streben] will not lead to mistakes."

By July he wrote: "My engine advances well. It will not take long to finish it. I have no doubts at all. If it is not this one, it will be another engine type. It must and will succeed."

The exceptional inventor was proving himself a most exceptional salesman. Having persuaded his preferred prospect to reverse a negative decision, Diesel moved effectively to sell rights and franchises for the projected but still unmaterialized compression engine. On February 21, 1893, he signed an agreement with the *Maschinenfabrik* Augsburg giving that manufacturing establishment exclusive rights for exploiting the proposed engine for southern Germany and sales rights for practically all of Germany. The company agreed to build and thoroughly test an engine within six months after the plans had been submitted by Diesel.

During the same month Diesel opened negotiations with the important Fried. Krupp *Werke* at Essen. On the following April 10, he signed a contract whereby the Krupp *Werke* received all rights to the engine for Germany not given to the Augsburg factory and the patent rights for Austria-Hungary. The contract stipulated that the Krupp *Werke* would also construct and test an engine of Diesel's specifications forthwith.

Next, Diesel corresponded with the Sulzer Brothers of Switzerland in an effort to persuade them to underwrite construction of the proposed engine. From his improvised laboratory at the Augsburg factory Diesel wrote Sulzer Brothers: "... It lies surely in the interest of all participants that only a few but excellent companies keep the matter in their hands and by common consent exclude all competition. In this manner a superb world business can be developed quietly."

The Sulzer company was gracious but chose to await the actual tests of the Augsburg model before undertaking manufacture. However, as solid evidence of interest, Sulzer Brothers offered to pay Diesel 20,000 marks (about $5,000) a year, pending completion of a practical model, for an option on Swiss patent rights.

When Sulzers hesitated, wanting to wait for early engine tests, and then made a rather hesitant offer for Swiss patents, Diesel wrote, "I have no interest in the amount of money as the situation now stands, but only in a contract with you." On May 16, 1893, the contract was signed. Because of the cost of the private laboratory work and the rising living standards of the Diesels, Diesel had found himself extremely short of money. He had, therefore, reversed his stand, signed the contract with Sulzers, and taken the cash.

As the Diesel talents for selling continued to speed ahead of the work developing his most aggressively exploited patents, his two chief customers, Krupp and the Augsburg factory, agreed to pool their efforts and share the expenses of constructing a pilot engine. The two firms joined in setting up a laboratory at the Augsburg plant, then plunged together into the formidable task of building a model engine. Their estimate engineers studied the problems of parts and drafted future plans for using machine tools for volume production of the various parts. The two companies then set out to build and test the proposed model. All agreed the "pilot" had to be impressively superior to any commercial engine then on the market. It was a big order. But Krupp and Augsburg were big *Maschinenfabriken.*

Lucian Vogel, the able Augsburg engineer who had been Diesel's student friend at Munich Polytechnic, joined in the work as inventor's aid. Vogel proved an energetic and dedicated helper. He began work with the well-implanted belief that the proposed engine would be successful.

Before the inventor returned to Augsburg in July, he wrote to his wife, who was then with the children in the country, "Tomorrow morning, Sunday, I ride to Augsburg for the most important and decisive moment of my life. Press your thumbs [cross your fingers] and pray for me."

On his return to his Augsburg laboratory, he found a letter from Martha: "My heart pounds when I think of your work and the difficult, difficult time of anticipation—Something may not work out right, I mean, some things may not be as ready as you expect them to be. That I am anxious and await your news, Dearest, you know, but I remain nicely content until you have time to be with your wife."

Rudolf Diesel arrived in Augsburg on July 17. The concrete foundation for the engine installation had already been poured.

Many parts of the engine were already set. The ten-foot-high black iron cylinder of the would-be engine was already erected. Most of the smaller parts were not yet in place, but they were being painstakingly tested.

Rudolf Diesel arrived with bags full of documents and an air of maturity. He was thirty-five. He had in hand enough drawings, diagrams, and pamphlets to fill half a dozen army trunks. This was only a beginning. Before the model building was completed, the sheaves of papers and stacks of notebooks were sufficient to fill a sizable library.

From the beginning of the Augsburg experiments, only liquid fuels were considered. Contrary to widespread popular belief, no tests for using coal dust were made during the first series of tests. The fuel oil first chosen for the tests was a tar-like *Pechelbrenner* —raw oil. This unrefined, very heavy hydrocarbon was a thick, brown mass which would not flow through the slender fuel lines at ordinary temperatures. Finding the heavy oil almost as difficult to ignite as it was to move, the model builders voted to abandon it. Rather than use valuable time trying to solve the obstinate fuel problem, they agreed to substitute gasoline and lamp kerosene for first tests and to concentrate all efforts on producing a workable model. Obviously, once an engine was built and functioning the selection of usable fuels would not be too difficult.

But an air-compressor device had to be installed to fill the supply tank. For this a one-stage ammonia compressor designed by Linde was used. At first external power was used to turn the engine until it would become possible for the engine to reverse the process.

The first engine tested (during July, 1893) was an upright single-cylinder machine. The oversize cylinder was mounted on an A-frame and had a 150-millimeter (5.91-inch) bore and a 400-millimeter (15.75-inch) stroke. The upper portion of the cylinder and the cover were manufactured from cast steel, while cast iron was used for the actual cylinder walls. There was no provision for cooling. The top of the piston was rounded, and the piston rod was connected to the large flywheel set into a well below the base line of the machine. The engine operated on the four-stroke principle.

The piston, made of cast steel, was to be kept tight and leak-proof by an oil pressure device employing thin, flexible bronze U-shaped rings, which, however, proved impractical and had to be abandoned. Several cast rings without tension were substituted and eventually were made operable, but not until after hundreds of trials with all kinds of materials and placement designs had been made.

The plunger of the fuel pump and needle was originally equipped with asbestos and, later, with leather. In subsequent tests both materials proved to be unsatisfactory. In contact with lamp kerosene the asbestos softened and the leather hardened. Next the leather bushings of the main valves burned off and had to be changed to a more durable material. Copper was tried, but asbestos was finally used. A still more vexatious problem was that of sealing the pressure air lines and valves. The petcocks, which were used to effect simplicity of control, could not withstand the great pressure. They simply had not been made for that kind of use.

In the first engine, the inlet valve and the exhaust valve were combined in a single unit, but in time this double-purpose valve had to be redesigned to allow satisfactory operation. The fuel-metering needle originally placed on the side near the operating gear was later moved to the top. Fuel injection took place directly from the pressurized fuel line, controlled by the timing cam. A spray nozzle was placed at the end of the long line in the cylinder itself.

When all these and other changes were executed, the beginning model operated effortlessly at three hundred revolutions per minute. At first, compression tests seemed successful only at eighteen atmospheres. The diagrams showed a great "negative area" and substantial power losses. After much further experimentation on the piston and valves, the compression rose eventually to twenty-one to twenty-two atmospheres (about 315 psig.) and, finally, to thirty-three to thirty-four atmospheres (about 485 psig.). This figure was, however, reached only after some twenty days of testing and experimenting. The theoretical pressure had been a figure twice as high. Vogel voted to let well enough alone.

Diesel launched a painstaking review of all available data and a re-examination of all parts which had an effect on compression. After diligent studies of the correct shape, size, and location of the combustion chamber in the cylinder, he and his colleagues double-checked the devices for atomizing the fuel particles. With meticulous, Germanic thoroughness the group prepared hundreds of diagrams showing the various curves resulting from the laborious and almost numberless tests and calculations.

The compression chamber of the first engine had been located like a convex scoop in the top of the curved piston; it covered an enclosure of 255 cubic centimeters (155.55 cubic inches). At that point a closer inspection revealed that other small irregularities produced an additional 157 cubic centimeters (95.77 cubic inches) of piston displacement space—an increase of more than 60 per cent over the original design. For obvious reasons this upset the first estimate substantially.

Despite the relatively low compression attained, the work crew proceeded to make a first test of the engine using the fuel injector. They used external power to position the big piston, then sprayed in gasoline vapor. On that hot August 10, Rudolf Diesel and Lucian Vogel waited anxiously to see if the contraption would run. The rest of the work force shared their anxiety. Then, to the joyful astonishment of most of those present, the engine ran properly. Ignition was prompt. The first indicator diagram showed compression up to about eighty atmospheres (1,160 psig.). Actually, the pressure may have been considerably greater. But that will never be known.

The indicator exploded. A bombardment of shattered glass and metal barely missed the heads of the two most intensely interested spectators, the Herren Diesel and Vogel. Luckily, the engine, built to withstand unusually high pressures, remained unharmed.

The following day the tests were continued, and again the engine ran with a regular rhythm, but occasionally loud explosions interrupted its regular pulsation. Thick, black clouds of smoke shot out of the exhaust, and in a short time the entire engine was thickly covered with soot. Valves and pistons blew mightily.

The first diagrams showed hardly any work performance, but eventually an increase was noted. The engine had now developed about 2.15 horsepower. Although the engine did not produce sufficient power to sustain its own operation, valuable observations had been made and important lessons had been learned. A detailed report on the thirty-eight days of testing operations was compiled. As always, the original of the report remained in the possession of Rudolf Diesel, who wrote all of the notes and made the diagrams. He dutifully delivered copies to the Augsburg *Maschinenfabrik* and the Fried. Krupp *Werke*.

Director Heinrich Buz wrote the following comment under his copy of the report: "The practical application of the process is proven in this imperfect engine."

Although its many components had worked quite well and almost exactly as anticipated, Rudolf Diesel realized that the engine in its present form could not in honesty be termed successful. It could not even power itself.

He returned to his home in Berlin to redesign the engine. On arrival in the Prussian capital, Rudolf deftly readjusted overhead to the situation. As a first step, the Diesels moved to a less costly apartment at the Brückenallee 15 and later from there to the Kantstrasse 153 in suburban Charlottenburg. There, Diesel's assisting engineer, Johannes Nadrowski, set to work on a drawing table in one of the rooms of the apartment.

The head of household, meanwhile, took time to attend to several long-delayed chores and privileges. He drafted applications for patents for his engine in a long list of foreign countries. Furthermore, he began to reacquaint himself with his own family.

Martha, at thirty-three, seemed more beautiful than on the evening he first met her. The three youngest Diesels were beginning to grow up. Even little Eugen was beginning to read. Hedy was, as her father noted, a quite sizable wisp, almost as pretty as she was energetic. "Little" Rudolf was now big enough to take part in the family's newest productive enterprise, which in peaceful little Charlottenburg was rabbit raising.

With systematic thoroughness Diesel set about building an array of rabbit hutches, in answer to an already materialized population explosion. When that was finished, he played with his offspring and with them devised special facilities for play. For a playroom he built a complete set of miniature furniture, each piece beautifully finished and painted. He then converted another room into a Chinese shadow theater. This, too, he designed, and he personally made the silhouettes. They had great charm and real beauty.

Daughter Hedy's verdict was, *"Wunderbar!"* R. Diesel chose to apply that same adjective to his young daughter, her older and younger brothers, and his pretty wife, Martha.

Back to Work

BERLIN WAS IN THE THROES of a political dispute. The young Kaiser, Wilhelm II, and the aging "Iron Chancellor," Otto von Bismarck, could not agree on policy. Diesel's sympathies seemed to be with the young emperor rather than the aging and battleprone count. Bismarck came out second best. Following his resignation, the national election yielded eye-opening results. The Socialist party abruptly emerged as a force of sufficient might to influence strongly the balance of political power. More than one and one-half million Socialist votes were counted in the national election. Anybody who thought that Karl Marx lacked actual, countable followers was thenceforward duty bound to think again.

Rudolf Diesel was pondering his own economic goals. He desired wealth that would enable him to complete his great invention and remain independent. In the progress of his inventions he would find a solution for the formidable social problems of the times.

After an exceptionally happy Christmas and New Year's at his home in Berlin, Diesel returned to Augsburg. There his sister Emma and her husband had built a home high over a narrow canal. A section of the house wall had once been a part of the ancient city wall. The winter-spring couple extended a customarily cordial invitation for Rudolf to live with them. Once again Diesel found himself a welcome guest of the Barnickels.

And once again he settled down to work.

Tests with the second model engine began almost at once. For the new effort the A-frame of the earlier design was retained *in toto*. As before, the dimensions of the upright cylinder were 150 millimeters (5.91 inches) bore and 400 millimeters (15.75 inches) stroke. But this time the upper cylinder chamber, cast from steel, was entirely redesigned. The inlet and outlet valves were rearranged and separated. The needle valve was placed in a removable housing on top of the upper cylinder chamber and immediately adjacent to the entrance of the fuel line into the combustion chamber.

The large exhaust valve had a smaller valve in its spindle which assisted in the action. The inlet valve was designed to serve doubly as safety valve, and control of a spring was provided by a small hand wheel above the inlet. The timing mechanism of the valves consisted, as before, of long rods which were connected with the timing mechanism at the base of the machine where the fuel pump was placed and where, on the earlier design, the fuel-metering needle was also located.

The piston, as promised, was completely redesigned. Its lubrication was achieved by means of an oil ring which would touch an oil sump at its lowest dead-center point. The combustion chamber was changed into a cup-shaped separate chamber in the cast iron piston and could be increased or decreased in size by adding easily removable inserts. No governor was provided for the engine.

The final testing period for the second engine began in January, 1894. Most of the comprehensive tests were made by Rudolf Diesel himself. A helper, Hans Linder, who had worked on refrigeration engines, and later a mechanic, Friedrich Schmucker, assisted in the tedious work. These two men, working alternately with Diesel, became intimately familiar with the various problems and characteristics of the engine which they helped nurse almost from infancy to final maturity. As a result of this immensely valuable experience, the two became the first competent engineers to make installations of the Diesel engines in many localities for the Augsburg manufacturing company.

As a foundation step, all important parts were subjected to extreme hydrostatic pressures; for example, the cylinder to 200 atmospheres, the cylinder head to 110 atmospheres, and the fuel injector to 160 atmospheres. This procedure was considered so important that it was later made part of the regular manufacturing process.

The engine was again started through the transmission by its small supplementary engine.

The valve springs were tested and adjusted by means of various weights, and their proper tolerances were carefully noted. Every change in the fuel-injection needle gave, of course, different results, and all were recorded with greatest care. The piston underwent minute examinations, and diagrams were made of the compression to determine the reason for compression losses. Since the piston rings on the first engine had been found to be too loose, Diesel designed new rings. These, however, proved so tight that they exerted a braking effect, a condition which showed up during heat tests on the cylinder wall. Various piston ring arrangements were then tried, two or three pairs of rings being used, with or without springs, or in variously spaced combinations. The compression pressure increased considerably and went as high as forty-four atmospheres. This was a definite step ahead in the desired improvement of the compression, but still not as high as Diesel believed theoretically possible.

Next came a radical change in the shape of the piston top. Because it proved impossible to prevent the rings from slipping off the cone-shaped head, this head was replaced by a flat center top. The prevention of leakage under high pressure of the various lines still posed a problem. All likely materials procurable were tested. Pressure of the various fuels tested—air, gasoline, lamp kerosene, and heavy oils—differed greatly and, consequently, presented diverse and, in many instances, annoying problems.

The cylinder head of the engine was removed, as on the first engine, to test again the formation of the spray in outside pure air; this at normal operation of the valve mechanism, as driven by the auxiliary engine. As a first step the spray was tested directly through the fuel-injection pump placed over an automatically controlled back-pressure valve located in the cylinder head. In this way a dependable spray of cone-shaped form and satisfactory density was achieved. At higher pressures the spray appeared fine, but at lower pressures a solid stream formed without vaporization.

Tests were then made of a procedure for injecting the fuel directly from the pump, simultaneously opening the metering valve and controlling the action of the piston and injection needle through the timing mechanism. The first use of the controlled injection needle in its later form was tested on January 30 (1894). The quantities of fuel admitted were metered by varying the length of the pump piston stroke. For this, various-sized connecting rods were used. The surplus fuel delivered by the pump was bypassed into the fuel line through valves which the piston held open.

The difficulties in developing a fuel pump to supply the minute amounts of fuel required in such a short time and at such high pressures still seemed insurmountable. Diesel decided to try making the injection of the fuel directly from the fuel line (which was kept under constant pressure) through the timing of the metering needle. This venture failed. Properly measuring the constant quantities of fuel at varying engine speeds seemed impossible. After prolonged tests Diesel decided to abandon this system because of the erratic behavior of the metering piston.

It was February before the inventor and his helpers could complete preparations for new tests. Working with able counselors, Diesel had by then provided the fuel-injection nozzle with a balanced drop valve. To that the pressurized fuel flowed and was controlled by the differential pressure of the incoming air. Small amounts of fuel were fed into this valve. Air pressure forced it through a measuring spool equipped with a multitude of small-diameter holes which served to atomize the fuel into a fine, cloudlike mist. The degree of air pressure controlled the density of the fuel cloud.

Experiments blowing the fuel into the cylinder of the engine resulted in strong explosions at top dead center and in forcing the safety valve to open at forty-eight atmospheres. The combustion attained was considerably quieter after the needle-timing mechanism was changed.

However brilliant the technical additions—and for the times, most of them were brilliant indeed—the engine still could not run on its own power. Diesel was certain it would, and that very soon. In that conviction the sharp-eyed men of Krupp and the Augsburg *Maschinenfabrik* remained almost as steadfast.

Abruptly and rewardingly, on February 17, 1894, the engine ran on its own power for the first time, at approximately eighty-eight revolutions per minute. Although the self-propulsion lasted only one minute, that minute was history-making.

When mechanic Linder observed that the belt which connected the flywheel of the Diesel engine with that of the auxiliary engine or "starter" jerked repeatedly, thereby telling that the Diesel engine pulled it, Linder lifted his cap slightly in a silent salute. So it was that Rudolf Diesel, noticing his assistant's gesture, realized that the Diesel engine actually ran. The inventor reached out his hand and clasped Linder's. Neither spoke; for the time words were wholly unnecessary.

Understandably, on that otherwise bleak February day in 1894, Diesel's hopes soared. He dispatched a letter asking his wife to come to Augsburg and share the triumph. Smiling and petite Martha came into the laboratory with her husband. She raised the lever which started the engine. Again the towering cylinder engine with the oversize flywheel ran smoothly. At long last "Rudolf's black mistress" was proving nobly faithful.

Stock market reactions were almost instant. The traded stock of the *Maschinenfabrik* Augsburg soared. Within a month the *Münchener Allgemeine Zeitung* reported (on March 17, 1894) that the *Maschinenfabrik's* stock had risen 30 per cent in lively trading, all because of the newly invented engine which the corporation had secured. It promised noteworthy successes. Diesel was, in his own words, *"ein Glückspilz* [a lucky mushroom]."

In April the inventor again traveled to France to meet his old friend, Frédéric Dyckhoff. The two walked the countryside of Bar-le-Duc. The stated purpose was to locate a site for a factory for building Diesel engines to distribute in France. But it was also April in Paris. Chestnuts were blooming. Rudolf was light-hearted. He sang *Lieder* of Schubert and ditties of the Paris streets. That evening for the first time in many months his long fingers moved over the piano keys. April in Paris was sweet. So was success.

In Paris Diesel met many influential industrialists. For a change all seemed interested in his work; several were interested in building his engine. He discussed the appearances and prospects with the *Société des Forges et Chantiers de la Méditerranée.* En route home he stopped at the Carels Frères in Ghent and at Cockerill in Liège and talked, too, with industrialists in Mül-hausen. The Carels Brothers paid him twenty thousand francs for the rights to manufacture Diesel engines for sale in Belgium and made plans for building a model at their own expense.

The news from Augsburg was less roseate, however. Further experiments at the factory with the now operating engine brought out that the problem of properly controlled combustion was still far from being solved. Diesel joined in experimenting with changes in shape and workings of the injection nozzle but to no avail. Back pressure still hindered the operation of the fuel-injection nozzle. The inventor and his helpers rebuilt the back-pressure valve and the drop valve.

The correct regulation and control of combustion still was not attained. Diesel by then was convinced that for reasons of safety the high-pressure fuel system should not have been con-nected so closely to the engine. A more practical system for supplying the exact amount of fuel necessary for one combustion cycle had yet to be found. Diesel and his counselors undertook trying and testing other means for blowing the fuel into the combustion chamber.

Then came the task of reproducing the precise conditions under which the engine had first operated successfully. With the throttle wide open, the model engine ran by itself for thirty-six minutes. During this period the amount of fuel was controlled by the drop valve. Diagrams were made to show the data for various load conditions. Most of these indicated unstable combustion—usually delayed ignition followed by detonations occurring rapidly one after another. During periods of smooth operation the output was equal to 13.2 horsepower at three hundred revolutions per minute, somewhere near the expected figures. But indicated pressure was 4.39 kilograms (9.66 pounds) a square centimeter. This suggested severe friction losses. To Diesel's profound relief a revision of the timing mechanism solved the problem of the delayed ignition periods. Some pre-ignition still occurred, but the mechanism was eventually corrected and operated in a satisfactory manner.

The forced injection of compressed air, either by an attached mechanism or by a separate compressor, appeared to be useful. Until this time a Linde compressor had been used in all of the tests made. It was imperative, however, that a workable method be found to insure proper metering of the fuel which was to be injected with and by way of compressed air. It was also evident that the compressed air had to be cooled, preferably in a long tube, before it came in contact with the fuel in the injection nozzle.

Since all systems for the injection of the fuel in its liquid form had the disadvantage of requiring too much time for atomizing the fuel (which resulted in low heat efficiency and excessive carbon deposits), Diesel and his helpers concluded that all known disadvantages would be overcome if the fuel could be injected in the form of vapor.

At first it appeared impossible to accomplish this "atomizing" in a satisfactory manner. There followed a systematic six months' search to solve the difficult problem of developing a mechanism which would convert the fuel to gaseous form, ignite it, and, at the same time, inject it properly into the heat chamber.

Earlier a type of carburetor had been built which was substantially a small kettle in which kerosene was heated by means of a burner placed beneath it. The heated gaseous fuel was then directed under pressure from the kettle into the injection nozzle. The basic theory seemed good, but in practice the process proved a failure. The gases cooled too much, and sufficient pressure could not be maintained at declining heat levels.

Stubbornly but carefully Diesel set out to build a completely new apparatus which fitted inside the chamber and required a redesigned cylinder head. This was made of cast iron. Next he installed a water-cooling device. This accomplishment involved passing the liquid fuel through a spiral tube prior to its entry into the nozzle. Then Diesel provided a manual control at the junction of the fuel line and the tubing. For testing the device, lamp kerosene was used exclusively. Occasional heavy knocks resulted, but they showed no regular pattern. Billowy white clouds of unburned kerosene vapor piled out of the exhaust. Even with the compression increased to 38 atmospheres, or some 540 pounds a square inch, the malfunction did not improve.

At the time, the cause of the difficulty was simply not known, and nobody seemed able to correct it. Diesel decided to institute a regular means of auxiliary ignition. He designed and built a mechanism which consisted of an asbestos wick supplied with kerosene by a small drip valve and ignited by a magneto. This, too, proved to be impractical.

A device, built by von Zettler of Munich, was then tested hopefully. The purpose of this invention was to effect ignition from outside the cylinder by generating electric sparks which "jumped" inside the cylinder between two stationary points. However, accumulations of heavy carbon deposits brought about short-circuiting.

Nearly three months of frantic work had resulted in failure and blasting disappointment. Haunted by a formidable string of failures, Diesel journeyed to Stuttgart to see Robert Bosch in the hope of securing a workable ignition system. Bosch assisted Diesel in completing and testing an apparatus in which the electric spark was to ignite the injected fuel and by which the fuel was to be supplied. However, the wick did not ignite either kerosene or gasoline. For that and other reasons, this device, too, proved useless.

Dyckhoff had constructed an engine in France which apparently functioned better than the one in Augsburg, and Diesel wanted to bring the French machine to Germany for study. He did not succeed in doing so.

Frustrating weeks followed. After six months Rudolf Diesel wrote his first two corporate sponsors, in a letter that was dated October 4, 1894, freely admitting his lack of success in solving the problem. He suggested that means should be developed for converting the fuel to gas—outside the cylinder.

Diesel wrote to his wife: "I hope that you, sweet wife, will remain true in your help and not give up. When you believe in me and my device I have the strength to work, otherwise not . . . ."

The hard-driven inventor's next move was to build a new engine which would use illuminating gas exclusively as fuel and then, with the knowledge gained from the performance of that fuel, go back to experimenting with liquid fuel. For practical use, illuminating gas remained almost prohibitively expensive, a fact that had thwarted the promise and progress of many otherwise creditable engines.

Despite the reverses suffered during this decidedly critical experimentation period, the directors of the companies—Heinrich Buz, Albert Schmitz, Ludwig Klüpfel, and Gisbert Gillhausen—stated their willingness to underwrite the continuation of experiments as proposed by Diesel. Rarely had a matter-of-fact industrial concern given a more devoted vote of confidence.

So began a period of experiments with illuminating gas. A city gas line, complete with meter, was laid directly to the laboratory. The one-stage compressor proved to be inadequate and was rebuilt into a multi-stage compressor of greater output.

At the same time, Diesel and his helpers built a new separate carburetor from readily available parts. The air which was blown into the cylinder of the engine went through a safety cylinder into a pineapple-shaped cylinder where it came in contact with the gaseous fuel which had been heated by the burner below. This flammable mixture then traveled through a longer line into the injection nozzle on top of the cylinder of the engine. The mechanism worked satisfactorily, and good diagrams of compression, admission, and expansion resulted. To his great relief Diesel found that the electric ignition as here provided was actually unnecessary. Compression alone was sufficient to effect the ignition.

A conference of the directors of the two companies (Augsburg and Krupp) was called, and the new development discussed. The verdict, however, was to go ahead as planned with illuminating gas. Convinced that the work with the liquid fuel showed promise, Diesel arranged the new machine so that it could also be powered with liquid fuel. A special test using gasoline showed a diagram similar to one of eight months earlier when a separate apparatus for turning the liquid fuel into gas was employed. Now the same results were obtained without such a device. Dependable heat was being provided by consistent pressure of confined air. The age-old principle of the blowgun was being updated for a brilliant if at times bumbling power era. Thus it became evident to Diesel that the proper mixing of the fuel with the air at the time of injection and the injection with air were the most important factors.

The correct mixture burned cleanly and developed the necessary energy which the compressed-air method had never been able to do alone. Diesel also observed that a burning fuel stream would be extinguished when too much air pressure was used. A weak or, in fact, an overly strong stream of liquid fuel did not ignite on contact even with red-hot iron. Instead, it cooled the iron and deposited soot. Experiments to discover a process for igniting the gas stream with electric sparks in the open chamber were also unsuccessful. Many other experiments for igniting gas followed. But trials proved that the consistent injection of a few drops of gasoline into the chamber ahead of the onrushing gas stream produced perfect ignition.

Comparisons of gasoline and kerosene fuel diagrams showed

that better results were obtained by using kerosene. This was due, perhaps, to the fact that sufficient air was being admitted along with that fuel. The necessity of admitting air was evident. Eventually, good results were obtained with illuminating gas, though several changes in the injection mechanism were necessary first.

Confidently Diesel concluded the ignition tests and drafted a detailed report of the findings. He firmly recommended that the actual combustion chamber be removed from the piston.

Austrian patent law directed that a successful machine had to be demonstrated within a prescribed period of time in order to confirm prior patent grant based on theoretical proposals. In the case of the new Diesel engine, this time limit had almost expired. Although the present engine was undependable, it functioned sufficiently well to be used as a demonstrator. Accordingly it was shipped to the *Metallwarenfabrik* Berndorf, a Krupp plant, for assembly. The metal works factory constructed various parts for the engine which incorporated the latest refinements.

Rudolf Diesel accompanied his invention to Austria. The engine was duly demonstrated before the patent commission in Vienna on January 17, 1895. Once more fate in Vienna was kind. The patent was granted. It was an important confirmation of "practicality" to come.

Not long after (July 3, 1895), Diesel wrote to his wife of the success: "My dearest wife, you shall still have a good-bye greeting, with the assurance that I love you, that I long for you, and that as soon as I can I will follow you. My engine still makes great progress; I am now so far beyond everything that has been accomplished until now that I can say I am, in this first and most elegant branch of the technic, the building of engines, the first on our small globe of earth, the leader of the entire column on this and the other side of the ocean. Are you not filled with pride at those words? I would almost become proud myself, if I were so inclined. But I am quietly happy inside, without song and gong, satisfied in the knowledge that I have accomplished a useful task and happy that our future is secure . . . ."

But Rudolf Diesel was mistaken.

An Engine Is an Engine Is an Engine

☙
☙

☙ BACK IN AUGSBURG, Diesel and his assistants set about building a refined and revised model of the engine. It was completed in 1895. Diesel was then in his thirty-seventh year and feeling every day of his age. The refurbished machine used the same base and gear as the previous model, but the bore was increased to 220 millimeters (8.66 inches) to double the diameter. The stroke remained at 400 millimeters (15.75 inches). The injection nozzle was located in the cylinder head, thus concentrating the compressed air there. Because of space limitations, the intake and exhaust valves were again combined as in the first model, but they were operated by a disk which opened and closed them alternately.

The cylinder was encased in a water jacket with the metering mechanism located on top to eliminate the usually long rods. Diesel had come to regard the ratio of the stroke to the cylinder diameter as of critical importance. The ratio on the first engine had been 2.67 to 1. On this new engine it was 1.82 to 1. Originally the piston diameter had been small to lessen its contribution to the weight of the assembled machine.

The next step was to develop a small, uniform combustion chamber. To insure ignition and for experimentation, Diesel and his men installed a spark plug sideways in the combustion chamber. They redesigned the needle housing to accommodate various test equipment. They provided two passages each equipped with fine-spray nozzles. The upper passage was for gas and air; the lower one for liquid fuel. The housing was made hollow to allow the pre-heating of fuel inside it. The piston was open at the bottom and was not water-cooled; instead it had expanders placed behind three sets of rings.

Building the new model provided valuable information regarding structural needs. For example, the first two cylinders cast were found to be porous; a correct technique for the required metallurgy had to be developed.

On March 30, 1895, a new patent, Number 86,633, covering the method of starting the engine, was issued by the *Reichspatentamt*. A later one, Number 90,544, of January 18, 1896, covered correlated phases.

The new tests began on March 26, 1895. A young helper, Fritz Reichenbach, a brother-in-law of Herrn Buz, took over as Diesel's assistant. They made ignition tests with the new spray nozzle which they had installed in the hope of securing a better spray pattern. The nozzle consisted of two star-shaped jets with many small holes all supplied from one centrally located tube. The location of the nozzle was so arranged that all of the air from the compression chamber had to pass over the jets.

Before more detailed tests were undertaken, accurate measurements of performances using the revised design showed that only 10 per cent of work in the combustion chamber space was lost, as against 28 per cent of the previous one and 60 per cent of the first engine. This improved proportion was expected to contribute to much better combustion.

After a month of tests with kerosene, gasoline was used; it was blown in with air by means of the double jets. The exhaust was smokeless and invisible. But the combustion was not instantaneous with the injection of the fuel. Only the mixing and conversion to gas ("gasifying") took place at the time of the injection. Tests indicated that the engine developed twenty-three horsepower at two hundred revolutions per minute, and thirty-four horsepower at three hundred revolutions per minute. This represented mechanical efficiency of 58 per cent.

Tests using lamp kerosene showed similar results but produced quieter operation. Use of illuminating gas—preceded by gasoline drops—also had the same effect, but using gas alone caused 50 per cent misfiring, owing to the lack of sufficient air. On May 1, 1895, about two years after the first practical experiments were begun, the anticipated performance was finally realized. For Diesel they had been great and terrible years.

Diesel next devised a friction brake. Because of greater piston pressure, the crankshaft overheated; as a result, frequent breakdowns occurred. Rather than rebuild the engine, the experimenters bored out the crankshaft housing and tried using water to cool it. Diesel then directed the designing of a new mechanism to atomize the fuel so that the "fogging" could take place in front of the spray nozzle.

Different tests, however, showed that atomization inside the spray nozzle was more satisfactory. Other tests using variously designed jets advanced the work. Star jets with suction rings and pipe jets with and without spiral tubing underwent countless tests with illuminating gas, then with injection of lamp kerosene, both directly and by means of the pre-injection atomizer.

On June 26 (1895), Diesel completed the first brake tests of the engine, using the "double-star" jets. Air was injected with the Linde compressor. Results were: internal thermal efficiency, 30.8 per cent; mechanical efficiency, 54 per cent; and brake thermal efficiency, 16.6 per cent. Even if the inventor had been the dancing sort, which he wasn't, these results were nothing to dance in the streets about.

Because of the low mechanical efficiency, the individual mechanical parts were examined. Improved lubrication of the piston and rings resulted in greatly increased efficiency. Diesel's meticulous calculation showed a rise in mechanical efficiency to 67.2 per cent; productive efficiency to 20.26 per cent. Further close examination showed that the cylinder was not perfectly round. There was also evidence that improved machine tools for the production of the cylinder would have to be constructed. Greater care was imperative.

In a letter to the two companies, Augsburg and Krupp, dated

July 8, 1895, Diesel compared his engine with a kerosene engine designed by Wilhelm Hartmann which Hartmann had just tested. Diesel's engine used only 60 per cent of the fuel required by Hartmann's. The power output of the two engines was equal, even though the piston displacement of Diesel's engine was only 58 per cent as large as that of Hartmann's. Furthermore, as of that time, the Diesel engine operated at only 75 per cent of its rated capacity.

Although the air needed for the revised Diesel engine was supplied by a separate compressor, the inventor began work to build and perfect an air pump. Also required were accurate gauges for the various measurable pressures. When the new gauges were built and installed, the tests showed higher fuel consumption and lower power output than had previously been attained without the pump. In any case, an independent engine —that is, without separate compressor—had been created.

Diesel centered all efforts on increasing the engine's horse-power output. He devised and directed continuous performance tests over long periods of time to gather pertinent engine information. The workers used both air pump and fuel pump. The first definitive findings showed that the engine remained clean, but the double-star jets lasted only approximately fifty hours in continuous operation. After that the carbon deposits curbed the engine's efficiency. Another weakness was that the jet could endure only a limited engine speed. Otherwise, the parts functioned quite well.

The feat of changing the jet was similar to replacing the carbon rod in an arc lamp. For the expert or the deft the chore did not take much time or effort. But making the change simple for the eventual user was a problem which would have to be solved.

A chemical engineer, Hartenstein, from the Krupp Works at Essen, arrived to make a detailed analysis of the exhaust gases. His tests showed that by improved utilization of the existing air, better combustion could be achieved.

Tests also showed that generally the volume of air in the cylinder was not as great as expected. In addition the combination inlet and outlet valve did not operate properly because some of the exhaust gas returned with the fresh air stream. These findings reaffirmed the conclusions reached about the earlier engine design. At that time Diesel had developed the combination valve in order to be able to place the valves within the crowded cylinder head. Now it would seem that a cylindrical combustion chamber of the same height and diameter was necessary.

The new air pump worked well mechanically, but the lubricating oil repeatedly caught fire. Greater cooling of the incoming air was essential. Moreover, the fuel pump gave considerable trouble. The bronze piston showed excessive wear, and the steel piston which replaced it did not stand up much better. Eventually a harder steel piston without the leather packing, which had earlier been used, gave satisfactory service.

Because of the favorable results obtained during the continuous performance tests and the other trials, the two interested companies believed that it was time for concentrating on developing a commercially acceptable engine. A conference was called on February 20, 1896. The Fried. Krupp *Werke* was represented by the Herren Asthöwer, Klüpfel, Albert Schmitz, Gisbert Gillhausen, Klemperer, and Ebbs. The Herren Heinrich Buz and Lucian Vogel represented the Augsburg *Maschinenfabrik,* A.G. (*Aktiengesellschaft,* or a stock company); Rudolf Diesel, of course, was alertly present.

It was decided to forego further testing and concentrate all efforts on manufacturing drawings of a one-cylinder engine of 250-millimeter (9.84 inches) stroke and 400-millimeter (15.75 inches) bore (with a ratio of 1.6 to 1) and completing the drawings for the compound engine.

Diesel not only directed the work in the laboratory but, for good measure, managed the adjoining engineering office where the drawings were made. Among the several young engineers who worked with him was Immanuel Lauster, who later made important contributions to the further development of the Diesel engine. The Augsburg factory had hired the twenty-four-year-old engineer in January, 1896, to help with Diesel's work. When the young man reported for work the day after New Year's, Diesel, to test his ability, asked him to figure the weight of the flywheel. He wanted the result that night. The answer tallied exactly with Diesel's computation.

When work on the drawings of the new engine was nearly complete, Diesel and his counselors decided to change them so that air was drawn into the cylinder beneath the piston in two stages. The completed drawings for the new engine were delivered to the factories on April 30, 1896.

The basic model underwent long testing for use in actual factories, power and light, or water plants. Under the supervision of the machinist, Friedrich Schmucker, it was run during regular working hours. Every morning the test engine was started, run all day, and stopped in the evening. The star jet had to be replaced with a clean one daily. To eliminate the troublesome jet exchanges, Diesel and his men tested a new-style jet continuously for seventeen hours. The cone-shaped jet performed well; it used 50 per cent to 70 per cent less air than the star jet, and almost all the holes remained open. Tests with illuminating gas also showed excellent results. After seventeen days of operation without any incidents, the engine was considered reliable and dependable enough to require no further testing.

It was now established that the engine without any readjustment would perform as well with illuminating gas as with kerosene. The cone-shaped jet was changed somewhat, resulting in a definitely satisfactory performance. So, on September 7 (1896), the now well-proved engine was dismantled after several photographs were made and filed for posterity.

Meanwhile, work on the new 250/400 (9.84 inches by 15.75 inches) engine with fuel pump had progressed to a point which justified the manufacture of component parts. The parts were now ready for testing. The cylinder, cylinder head, and cooling jacket were built of cast iron. The combustion chamber was a one-unit, smooth chamber between piston and cylinder head without side chambers or bulges. All such space losses had been successfully eliminated. The piston was hollow and water-cooled. Four expander rings were used. Inlet and outlet valves were separated. The air and fuel nozzles entered the injector through the side of the cylinder head. The timing mechanism was placed on the cylinder head and connected with the fuel pump, which was moved near the injection nozzle. The bottom portion of the working cylinder was covered with a housing containing the inlet and outlet air valves for the pump. A patent, Number 95,680, was issued to Diesel on March 6, 1896, covering these adjustments.

The air-pump cylinder was an extension of the working cylinder. The one-piece air receiver (made of steel), which also served as injector bottle, had a safety valve similar to that on compressors of that era. The safety valve consisted of a rupture disk designed to burst at a predetermined pressure. The disk had been calibrated by water-pressure tests.

During the testing periods, safety valves were placed in all dangerous places, such as the injection mechanism, the outside carburetor, and the cylinder itself. Other precautions taken during the experiments included the installation of special gravel pots, or check valves, in the more dangerous lines to prevent backflash into the fuel supplies. The record of no accidents during the entire five years of experimentation was, and still remains, remarkable. But, significantly, the wall of the laboratory, which faced most of the safety disks, was well peppered with pockmarks. The Diesel engine was born dangerously but with memorable concern for safety.

The manufacture of the various engine parts was not without difficulties and challenges. First, cylinders and air-pump cover were made useless by porous spots in the metal. As a result precision casting and improvements in iron and steel work were sought.

Having survived the final two years of engine testing, Diesel welcomed the chance to move his family from Berlin to Munich in order to be within easy access of the Augsburg factory. He was determined to follow closely and personally the manufacturing of the remarkable new engines. With three young Diesels to humor, the family moved into Giselastrasse 14, in nearby Schwabing.

Once more the father had money to spend. Shortly before Christmas, 1895, the Diesels had invaded Munich for a shopping spree. They viewed and marveled at the uneven, helmeted towers of the *Frauenkirche.* The ancient Bavarian stronghold was losing its ancient markings, however; smoke-spilling stacks of great new manufacturing plants and breweries were evident.

The arc lights on Munich's Ludwigstrasse were sparked by electricity. There were still horse-drawn streetcars, but the line to Giesing was now fully electrified. Soon the web of electric power spread across the Bavarian countryside to hamlets and towns. The old king-city Munich was different from the bustling, Kaiser-city Berlin. Diesel thoroughly enjoyed Munich's museums, theaters, and concert halls. In them he found a resurgence of youth. With Martha he crowded into a performance of Richard Wagner's *Die Meistersinger von Nürnberg.* On first exposure he found the uproarious pageant completely delightful. He was no less delighted with it than he was with all the modern music of the still controversial German master. Once more Rudolf Diesel could revel in great music and machines.

By October 6, 1896, the first of his new engines was completed. Its manufacture had taken five months. Ten days later Diesel began supervising the comprehensive tests.

The piston lubrication had to be improved; this was accomplished by installing a new pressure system, which lubricated the piston in four places between the rings. The valves were tight and required adjustment to give them more play. The fuel pump did not function properly. Again, various materials were tried for the piston packing, such as leather, metal, and cotton. Another type of jet, one which scattered the fuel in a uniform pattern, was tested. The fuel was sprayed directly into the chamber by the jet. If proper regulation could be achieved in this manner, the fuel pot with its lines, petcocks, and other complicated mechanism could be eliminated. The sprayer was located now at the discharge end of the injector. The atomizer consisted of two horizontal disks which had a number of small holes between which a fine wire mesh was placed. The operation of the sprayer was most satisfactory.

On the last day of 1896 the engine was run with all parts duly corrected and improved. Brake tests on January 29, 1897, produced the following results: thermal efficiency, 31.9 per cent (38.4 per cent at half-load); mechanical efficiency, 75.6 per cent (61.5 per cent at half-load); and productive efficiency, 24.2 per cent (23.6 per cent at half-load.)

The operation of the engine was excellent. At long last, success seemed assured for Rudolf Diesel. Many agreed that what he termed "the prime mover" was by self-made evidence the century's most important invention.

Laurel for the Victor's Brow, Ashes for His Cheeks

WITH THE SUCCESS of his engine assured, Diesel contemplated moving to Augsburg. After thinking it over, however, he found that Munich appealed to him more. He wrote: "In Munich one has the fine, easy reach of surroundings, the tourist traffic, the museums, the wonderful art exhibits, the theater, the mental stimulation of the large city in which art and knowledge flower, and the possibilities of acquaintance with leading intellectuals." He had heard the *Meistersinger,* and wrote: ". . . an excellent work, that made an overwhelming impression on me, as a natural wonder of great importance, the mental greatness of Wagner is really almost incomprehensible. . . . I was almost half-dead from exhaustion afterwards, because I followed it all so intensely; five hours long."

In July, 1897, the inventor moved his family to the Schackstrasse 2 in Munich, where they occupied an attractive apartment on the third floor. The then substantial rent was some four hundred marks a month. Diesel now had real money to spend. He and his wife employed a French governess for the children; then they hired a butler; and, to complete the sequence, they commissioned a portrait painter to make oil paintings of both Rudolf and Martha. The fee for the paintings was eleven hundred marks each. They regularly went to concerts, opera, and the theater. Diesel became interested in photography, bought much equipment, and began to develop his own pictures. Money flowed freely, and close relatives were supported willingly. Papa

Theodor was no longer obliged to practice magnetism. The inventor's mother was well supplied with spending money. The new engine was the money raiser. The endearing old town of Augsburg was well aware that the Diesel engine had a very special appointment with history and industrial destiny. By February, 1896, the neat old city of historic landmarks was becoming a mecca for interested industrialists from points near and far. The knowledge that the development of dependable and efficient engines was an essential to the on-surging industrial age was companion to the realization that what was being called the mechanical revolution had decidedly come to stay. It was also accepted that the industrial revolution could not last indefinitely if powered by the brawn and backs of manual laborers. Even the stubborn English were granting this. Their new factory towns and suburbs were already grimy with coal smoke from the boilers of outrageously inefficient steam engines.

Better engines simply had to be developed, for smaller engines were needed by smaller shops and factories. According to the most respected appraisals, Diesel's was the answer to the obvious need for a new and better engine.

Among the first onrush of Augsburg's important visitors was Diesel's good friend Frédéric Dyckhoff from Paris. Dyckhoff had watched the February tests which showed the best results yet attained. Thermal efficiency ranged from 34 to 38 per cent, and productive efficiency was 26.6 per cent, both at full load. This was as good as the best gas engines were then able to produce, even under the most favorable circumstances. The gas engines used expensive illuminating gas as fuel; Diesel's engine was burning low-grade kerosene and could almost certainly be adapted to still cheaper fuels. Its output of usable power was approximately double that of other engines of similar size. Monsieur Dyckhoff was most favorably impressed. He telegraphed to his French associates the news of the "perfected" engine.

Three days later representatives of the *Gasmotorenfabrik* Deutz, at the time the leading European manufacturer of gas engines, came to examine the Diesel engine. Rudolf Diesel had been greatly interested in getting the huge Deutz plant to build his engines. But their investment in the Otto gas engines was so great, that they did not wish to take on a competing product even though it might be a better one.

At the next demonstration of the Diesel engine, the Krupp director, Gisbert Gillhausen, was present along with Director Schumm and Engineer Stein of the Deutz factory. The machine was put to many difficult tests. It was not permitted to warm up properly before brake tests were made with water temperature at merely 17 degrees centigrade (62.6 degrees Fahrenheit). The engine was accelerated to its maximum revolutions. When that point was reached, the supply of fuel was stopped abruptly and then started again quickly. The machine continued to function —better than most engines could under the best attainable conditions.

The representatives of the Deutz gas engine factory were so deeply impressed that they notified the Augsburg-Krupp consortium that they felt the Diesel patents could be questioned. They pointed out that before Diesel, such men as Otto Köhler, Emil Capitaine, Julius Söhnlein, and Gottlieb Daimler had either made experiments with, or received patents for, engines of similar principles. Having suggested patent infringement, the Deutz officials left the door open to an amiable settlement, so to speak. They offered 5 per cent royalties but no additional payment for the privilege of manufacturing Diesel's engine.

With poet-like diffidence Diesel ignored the offer and the insinuations. He was confident that he had not infringed any existing patents. Heinrich Buz, his manufacturing director at Augsburg, agreed with him. Diesel suggested that if Deutz proposed to fight his patent rights, the firm should not be allowed to manufacture his engines. He was certain that he and the Augsburg-Krupp combine could not lose in any legal assault on his folio of patents.

The point of view was far from unanimous. Professor Otto Köhler of the *Königlichen Maschinenbauschule* (Royal Machine

Building School) of Dortmund was also the advising engineer at the Deutz works. Köhler had published a paper which contained ideas similar to those of Diesel, though at that point nothing had been done to test the professor's theses. Even so, some experts believed that the claim against the priority of the Diesel patents was strong and that a court decision could have invalidated them.

As one scientist to another, Diesel met Köhler. The two promptly became fast friends. The threatened court action never materialized.

On July 19, 1897, the Deutz *Gasmotorenfabrik* signed a contract with the Augsburg-Krupp syndicate. The terms called for the immediate payment to Diesel of fifty thousand marks and 20 per cent to 30 per cent of the value, according to size, of each engine sold, with a minimum of twenty thousand marks. The Deutz company agreed to recognize the validity of the Diesel patents. That again was sweet wine to the inventor who imbibed but modestly and infrequently.

Four and one-half years had already passed since the granting of the basic patent. Since objections to status of patent rights had to be filed within five years, Diesel's position now seemed secure. But it was not.

For reasons not clearly explained the Krupp directors remained concerned about the validity of Diesel's patents. At the directors' conference of March 11, 1897, a suggestion was put forward that Diesel's engines *not* be manufactured. Rudolf Diesel was on hand at the Essen meeting, which raged on for eight hours. With Heinrich Buz to lend support, he eloquently convinced the wavering directors of the stability of his patents. Once more Buz had expressed complete confidence in the patent rights of the Diesel engine.

Following Krupp confirmation, a new contract was made directly with Diesel as inventor. Earlier he had waived his rights to any percentage license fees; instead, he had received thirty thousand marks annually during the time of construction of the engine. Now that the experiments had been concluded and a marketable engine had been created, it was agreed that Rudolf

Diesel was to receive fifty thousand marks yearly, still without percentage license fees, but with the absolute assurance that the Diesel engine would be manufactured and sold commercially.

Professor Schröter, who had indicated publicly five years earlier his belief in the proposed thermo-engine when only the theoretical calculations were made known, was asked to conduct an independent, exhaustive test on the finished engine and make a complete evaluation report to the Augsburg and Krupp directors.

The scholar stated in his lengthy, detailed findings that the Diesel engine, although not realizing fully all possible developments of the one-cylinder, four-stroke-cycle engine, stood easily at the head of all heat engines then known. At its rated output of eighteen to twenty horsepower, at normal engine speed and fueled with kerosene, the machine converted 26.2 per cent of the potential heat energy of the fuel to effective work. This was very far ahead of the proved record of any other engine then on the market. The mechanical efficiency at full load was 75 per cent, and thermal efficiency at full load 34.2 per cent; 38.4 per cent at half-load. The unusually simple method of regulation made it possible to change quickly from full load to any load by controlling the amount of fuel. This elasticity of operation, so cherished in steam engines, was available to a like degree in this internal-combustion engine. The ease with which it was possible to start the cold engine was especially important. Professor Schröter concluded that the engine, in its present form, would prove itself a thoroughly marketable machine. He called it the "engine of the future."

The inventor joined his associates to make further improvements. To allow for lighter gear, they increased the ratio of piston diameter to stroke to 1.5 to 1. After running the old test engine for 255 hours, a careful examination of the scatter jet showed that it remained clean and in operating order, although slight oxidation showed on the edges of the wire mesh. Eventually, the devoted testers eliminated the troublesome wire mesh and replaced it with two perforated disks.

During tests and demonstrations to representatives of interested parties, the engine developed a mechanical efficiency of up to 80.5 per cent; internal thermal efficiency of up to 38.7 per cent; and brake thermal efficiency of up to 30.2 per cent. Fuel consumption was 211 grams of kerosene per horsepower hour. These statistics were far ahead of what any competitor could prove. For that matter, they were beyond the ready capability of most power engines.

Improvements continued. Toward the end of May, 1897, the 90/200 (3.54 inches by 7.87 inches) air pump was replaced by a smaller unit of 40-millimeter bore and 60-millimeter stroke (1.58 inches by 2.36 inches). The air was no longer taken out of the atmosphere, but out of the air compression chamber of the working cylinder. This high-pressure pump was found adequate and was duly patented by the Augsburg factory.

The small laboratory had served well its appointed task. The proposed idea of an engine had been solidly embodied, and the fundamental laws and typical construction of the Diesel engine had been established. Sufficiently so that the manufacturers could now begin the building of the machines for commercial uses. After five active, immensely exciting, but often extremely frustrating years, the laboratory was closed. It was commonly accepted that the chief mission of the inventor Rudolf Diesel had been accomplished.

Quite clearly Rudolf Diesel was happy. But very shortly thereafter he again began wading the pitfalls and quicksands of the new modernity. For one dire step he acquired one of those American contraptions called a telephone and had it installed in his apartment at Munich. Shortly thereafter came the tremendous thrill of a first long-distance call—from a friend in faraway Berlin. The voice carried strong and clear. In practically no time the same wonder instrument began bringing news of tumbling stocks, engine difficulties, license disagreements, and a hundred other troubles. "So I just would have a telephone!" Diesel moaned.

The inventor also invested in another American machine,

Courtesy Deutsches Museum, Munich

Rudolf and Martha Flasche Diesel at the time of their marriage, 1883.

Courtesy Deutsches Museum, Mu

Diorama of the Diesel workshop in the *Maschinenfabrik* Augsburg, in
July, 1893, showing the first test engine.

Courtesy Fried. Krupp Werke

Munich exhibition (1898) display of the Diesel engines of the Fried.
Krupp *Werke, Gasmotorenfabrik* Deutz, *Maschinenfabrik* Augsburg,
and *Maschinenbau* A. G. Nürnberg.

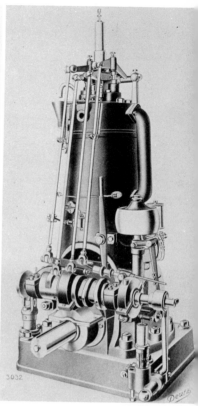

Courtesy Mirrlees, Bickerton & Day, Ltd.

Courtesy Deu

On the left, the first Diesel engine in Great Britain, built by Mirrlees, Watson & Yaryan Co. in 1897. Now in the South Kensington Science Museum. On the right, a Diesel engine, eliminating the crosshead, with direct air injection, built by Deutz Works, Cologne, in 1900.

Courtesy Busch-Sulzer

First Diesel engine to be placed in commercial service in the United States—a two-cylinder, 60-horsepower Diesel engine built in the United States and installed in 1898 at the Anheuser-Busch Brewery.

Courtesy Sautter, Harlé et

Courtesy Fried. Krupp W

Two Diesel marine engines. Above, a two-cylinder, horizontally opposed, engine of 25 horsepower developing 360 rpm, built in 1901 by Sautter Harlé *et Cie.*, Paris. Below, the six-cylinder, four-stroke-cycle engine, 450 horsepower at 400 rpm, built by the *Krupp-Germaniawerft* for the submarine *Deutschland* in 1913.

Courtesy M.A.N.

The first large stationary Diesel installation, consisting of six M.A.N.
four-cylinder units of 400 horsepower each,
at Kiev, Russia, 1903–1904.

Courtesy Busch-Su

Three-cylinder Type A Diesel engine, 75 horsepower at 250 rpm, built by the American Diesel Engine Company in 1907.

though this one was of German invention. It was called a type-writer. The contraption had an ink pad instead of a ribbon. It bore the trade name "Yost," and it not only inked decipherable letters, but it made more or less decipherable copies of letters. The weakness was that the machine required an "operator." Diesel had a secretary, but typewriting was just too much. He collapsed under the strain and became quite ill.

Diesel, too, met an unhappy fate. His violent headaches began to return. He found himself unable to relax and virtually un-able to sleep. The inventor turned to a chemist friend, Adolf von Baeyer, who was experimenting with an extraction from coal tar, a fever-reducing painkiller named antipyrin. Diesel began taking it. Presumably it temporarily relieved his headaches, but there was no evidence that antipyrin helped his rebelliously tense nerves.

When the headaches returned, Diesel visited one doctor after another. One physician blandly assured him that his mind was failing. Another, perhaps with a sadistic bent, burned out his nasal passages—presumably a trial-by-fire sinus treatment. The more rational of the medical plum pickers reiterated that as soon as Diesel's engine began functioning consistently, the migraine "condition" would vanish like a Wagnerian tenor heading for the *Biergarten*.

However perceptive and accurate, the deductive diagnosis was less than joyful to the pain-racked patient. The engine was improving, but there were problems. During April, 1897, Chief Engineer Lucian Vogel, a good and loyal friend of Diesel's, left the Augsburg factory. Vogel did not agree with the policy of speedy production of a marketable "Diesel." He urged more tests and greater caution than the Augsburg-Krupp group and, indeed, Diesel deemed necessary. The loss of Vogel was a major disappointment to the inventor. The two had worked closely and understandingly through the early, difficult days of pains-taking and sometimes painful experiments.

The new chief engineer, Josef Krumper, was a steam-engine expert who had occasionally visited the busy laboratory in which

Diesel worked. He had rarely left the premises without making a sarcastic remark to belittle the work. Understandably no particular love was lost between Diesel and Herrn Krumper. Fortunately the latter had as assistant an engineer named Hans Vogt who took charge of the engine-making. Vogt was not only sympathetic but enthusiastic about the new-principle engine.

The Augsburg factory promptly mailed a circular to their prospective customers, announcing that a Diesel engine developing twenty effective horsepower had been built and successfully tested over a period of several months. More or less simultaneously conferences for licensing agreements with the Augsburg-Krupp syndicate followed and were concluded on July 24, 1897. Shortly several other demonstrations resulted in license or franchise arrangements within Germany.

The Augsburg-Krupp combine held basic manufacturing and selling rights for the Diesel engine for Germany, but the rights for all of the other countries remained in Diesel's personal possession. He was free to make whatever arrangements he could and to receive whatever sums he was able to secure for the building and marketing of his engines from sources outside Germany. The still youthful-looking inventor continued to wait as sweet fruits fell into his lap.

At Augsburg the parade of interested visitors continued. In its ranks were technicians and industrial leaders from a dozen countries. The factory-minded world required a combustion engine of superior design.

Representatives of Mirrlees, Watson and Yaryan Company of Glasgow came trooping to Augsburg. Messrs. Watson, Robertson and Platt stayed for several days to study the engine which seemed to respond to Scot reverence for frugality. Even with an overload of 23 per cent, the engine performed well. After a forty-five-hour run the jet did not require cleaning.

Shortly thereafter, Rudolf Diesel went to London to confer with the directors of Mirrlees, Watson and Yaryan Company regarding license arrangements in Great Britain. The discussions between Diesel and the other men dragged on. Scotsman Robertson was particularly hard to deal with. A battery of solicitors sat behind him at the conference table. After a few days, the group adjourned to Glasgow for further discussions.

During his stay there, Diesel was a guest in the sumptuous Robertson home. He was immensely impressed with the glass-domed, electrically illuminated mansion. By comparison Diesel's own home seemed to him like a rude cabin. Later, when he visited the palatial home of another Scottish industrialist, Mirrlees, it was the Robertson home that seemed ordinary. The meals were served elegantly by liveried lackeys, and the table was exquisitely decorated with rare orchids. Only the food was meager by what Diesel termed Continental standards.

The 1897 meeting with the venerable Lord Kelvin, recognized as one of the greatest minds in the field of physics and heat theory who had proposed the thermodynamic temperature scale which bears his name, and their extensive walks in the spacious grounds of the university where the brilliant James Watt once had walked were high points of Diesel's visit.

The question of the validity of the Diesel patents, first raised by Deutz, had followed the inventor to Scotland. Upon the valued advice of Lord Kelvin, a licensing contract was finally signed on March 23, 1897, giving the Mirrlees company the privilege of manufacturing and marketing the Diesel engines in Great Britain. The Scots of Glasgow agreed to pay Diesel twenty thousand marks annually. The agreement marked the Glasgow company as the first foreign factory to undertake the building of a Diesel engine. A pilot model was completed within three months. Tests showed that the output was identical with the best results obtained in Augsburg.

One of Diesel's assistants, the engineer Anton Böttcher, super-vised the installation of this first British Diesel engine. Some fifteen years later when an inspection was made of the first British installation, the engine of 1897 was still performing well.

From Glasgow, Diesel traveled to Winterthur, Switzerland, to talk with the *Gebrüder* Sulzer Company. The Herren Sulzer-Imhoof and Sulzer-Schmidt, and engineer Erich Brown, representing the Sulzer Brothers, then the largest of Swiss machine builders, had also come to Augsburg in February to inspect the Diesel engine minutely. Now Jakob Sulzer-Imhoof, acting head of the plant, was agreeable to Diesel's suggestion of an agreement whereby Sulzers would also manufacture the engine. The following year the Swiss firm built its first model.

In April, 1897, a group of interested French industrialists founded the *Société Française des Moteurs Diesel* at Bar-le-Duc. Later that month a representative committee, consisting of Edouard Sauvage, professor of the *École des Mines,* Paul Carie, engineer of the *Société des Forges et Chantiers de la Méditerranée,* and the engineers Dyckhoff and Merceron, came to see the Diesel engine at work. Presently (in June of 1897), Diesel went to Bar-le-Duc, where the French company for building Diesel engines was then being formed with the enthusiastic Dyckhoff as its chief engineer. The inventor received stock in the new company valued at 600,000 francs.

Although engine demonstrations and contract negotiations were taking most of Diesel's time, the inventor began making lecture appearances before various interested groups. In accepting the Bavarian Michel medal, Diesel gave a detailed report of the construction and performance of his engine. For the occasion the canteen of the Augsburg factory was decorated gaily in bunting of blue and white, the colors of Bavaria. The audience included many of the local industrial leaders. The occasion marked the inventor's first public appearance in five years; apparently he enjoyed it.

Diesel began accepting other speaking engagements. In mid-June he explained the progress of the engine to the Bavarian District Union. Next day (June 16, 1897), he arrived at Kassel to address the main assembly of the Union of German Engineers. There the inventor offered a scholarly and fully detailed report, stressing the fact that his thermo-engine represented a necessary compromise of theory and practice. From a standpoint of public image the Kassel address was almost certainly the most important in the inventor's career.

Professor Schröter also addressed the assembly on "Diesel's *Rationeller Wärmemotor*" [Rational Heat-Engine], but Rudolf Diesel was the star of the seminar, and he knew it. Next day he wrote his wife jubilantly with his earlier and, at times, childish delight in juggling the German language, ". . . . The weather is superb here, real *Dieselwetter*. . . . Yesterday was a great day, a *wertlicher* [real], *aufrichtiger* [honest], and *durchschlagender* [decisive] success which probably will also bring many economic advantages. . . . All of the German high schools [colleges] were represented by one or more professors, and everybody who is anybody and has anything to do with engines in Germany was present."

But the elation was again short-lived. Rudolf Diesel was on the verge of an illness which was to result in a serious breakdown of health. Diesel confided to Ludwig Noé: "Can you spend a few years of your life on an idea? Can you tie yourself for a few years to a sick man? I cannot do anything any more, I am finished. Oh, my head, my head!"

CHAPTER ELEVEN

"Build a Better ═══"

WORD OF THE WONDER ENGINE was getting to the United States. To name just one instance, Bernhard Bing of Nürnberg, who dealt in and exported hops, had brought the Diesel invention to the attention of Adolphus Busch, the beer magnate of St. Louis, Missouri.

Herr Busch had followed the westerly beer trails with spectacular success. Before the Civil War, German-Americans had penetrated the New World and provided a market for the homeland *Braumeister*. During the quarter-century following the Civil War, the amount of beer consumption in the United States was no longer determined by the demands of the German segment of the population. The brewery strongholds, however, remained in the principal areas of German settlement, beginning with eastern Pennsylvania and extending to St. Louis.

The onetime Indian-French trading post on the middle Mississippi was an outpost of Germanic immigration and beer, both of merit. Of all the transplanted brewing clans, the Busches, formerly of Rheinhessen, were St. Louis's most successful and affluent. Thanks to Herrn Adolphus *und Kinder,* St. Louis by the 1890's was being designated as "Buschberg." This was more than passing flippancy, and Adolphus Busch had more to offer than pomposity. His first brewery complex had proved fabulously profitable. That beer-making brings large profits, A. Busch was demonstrating with girth and fervor.

But Herr Busch was interested in other productive enterprises.

Admittedly his first interest in the Diesel engine was for powering breweries. Beyond this, however, Busch was desirous of exploring and building other industries for his adopted American city which he saw as greater than a brewery park with *Biergarten*. Unlike certain competitors in, say, Milwaukee, Herr Adolphus went so far as to view with approval the teaching of English—American English—in St. Louis schools, instead of compulsory German. His many activities extended outside the brewery.

In September, 1897, in keeping with this point of view, Adolphus Busch dispatched his friend and technical adviser, Colonel Edward D. Meier (former U.S. Army engineer) to Augsburg to examine the Diesel engine and report in detail.

Directly thereafter, and one may guess by more than coincidence, Adolphus Busch hied forth for Germany, trailed by an entourage of some fifty friends, kin, and retainers. He led the way to the sumptuous resort center Baden-Baden, and there took an entire floor in one of the better hotels. St. Louis's pre-eminent brewer knew how to spend money as well as make it.

Colonel Meier, accompanied by Georg Marx, chief engineer of the famed *Maschinenbaugesellschaft* Nürnberg, had gone directly to the Augsburg *Maschinenfabrik*. The two had inspected the engine and completed an encyclopedic report, including minute data filling armloads of notebooks. For good measure the Colonel had managed to interview or pick the minds of other informed parties. These included Anton von Rieppel, well-known bridge builder and general director of the Nürnberg factory; Lucian Vogel, the chief engineer, and Heinrich Buz, director of the Augsburg factory; and engineer Schmucker, who had been close to the development of the engine. With Germanic thoroughness Meier devoted several pages to Diesel himself—his record in school, his writings on the principle of the engine, and his experimental work on the engine.

On October 4, 1897, Meier personally delivered his report.

The American engineer foresaw a glowing future for the Diesel engine. He saw no reason why it should not replace the steam engine completely, including its job of powering mighty warships and other ocean craft. Then there were automobiles. In all, the Colonel recommended urgently the acquisition of the American patent rights. Astutely, he extended a very cordial invitation to the inventor to come to Baden-Baden.

Diesel was growing accustomed to gracious invitations. At Baden-Baden he encountered what he could describe only as a *grossartige* reception. The great Adolphus Busch was hospitable, indeed jovial. His jacket pockets bulged with American-minted five- and ten-dollar gold pieces, with some twenties or double-eagles thrown in, for handing out.

The soft-spoken guest noted the unusual eagerness of the hotel staff to serve and accommodate the man from "Zaint Louee." To reward any solicitous service even as slight as cleaning an ash tray or refilling an empty ice bucket, Adolphus Busch handed out at least one gold piece. For more substantial service, such as opening a door or lifting a bag, he gave still larger gold pieces.

As Diesel drank in the lavishness of the great American success story, he apparently made some financial estimates of his own. When the conversation got around to a reasonable first fee for an American franchise, the inventor, perhaps with faltering breath, boldly named one million marks, then about a quarter of a million dollars, as an "anticipated fee" for licensing the manufacture of the Diesel engine in the United States.

Without hesitating or registering surprise Adolphus Busch accepted, wrote out a check, and directed that the contract be drawn up for signature. At least momentarily it appeared to be the climax of the up-and-down saga of Diesel's salesmanship.

The history-making inquiries continued. In October, 1897, representatives of both the British firm of Vickers Sons and Maxim Limited, and of the eminent Danish works of Burmeister and Wain of Copenhagen (the latter sent Professor Wilhelm Hartmann of Berlin and Professor Winslow of Copenhagen as consultants) made a thorough study of the Diesel engine. In November the Howaldt brothers came from the Baltic port city of Kiel to evaluate the engine's adaptability for marine purposes.

As things developed, *Motorschiff Monte Penedo,* the first German motor ship to be equipped with Diesel engines (in 1912), was supplied by Howaldt *(Gebrüder-Howaldt-Werft),* who used a Sulzers-made engine. The Diesel engine was destined for international development and service.

In November, 1897, *Maschinenfabrik* Augsburg delivered its first commercial Diesel engine to a customer; it was a 76-horsepower unit delivered to *Aktiengesellschaft* Union at Kempten. The engine was installed under the supervision of *Maschinist* Schmucker, with two American engineers—Meier and Puchta— as extremely interested onlookers. Fifteen years later when inquiries were made regarding Germany's first commercial installation, the venerable engine was still in excellent working condition. No major repairs except replacements as a result of regular wear had been necessary. Experts agreed that the engine was still good for many more years of satisfactory service.

On New Year's Day, 1898, a new manufacturing company to build Diesel engines was established in Augsburg. Forthrightly named *Die Dieselmotorenfabrik* Augsburg, it was underwritten by the bankers Schwarz and Gerstle of Augsburg. The capital investment was 1,500,000 marks, of which Diesel contributed 100,000 marks. Although never active in its management, the inventor remained a member of the company's board of directors. Since by 1898 the right to manufacture Diesel engines in Germany was held by the *Maschinenfabrik* Augsburg and the Krupp *Werke,* the new company was obliged to pay 100,000 marks for a license from the Augsburg-Krupp combine to manufacture the engine. "It is like buying cider from one's own apple tree," Diesel reflected.

Among the engineers who had heard Diesel speak at Kassel was Anton Carlsund, then in the employ of Emanuel Nobel, a nephew of the famed Swedish dynamite and explosives "king" and philanthropist. Emanuel Nobel owned large oil holdings in the Baku frontage of the Caspian, as well as a machine factory in St. Petersburg. Therefore, he had a double interest in the present and future of the Diesel engine.

While the inventor tarried at Baden-Baden, Nobel's engineer arrived at Augsburg with a Professor von Döpp. Shortly thereafter Diesel received an urgent invitation to call on Nobel at the Bristol Hotel in Berlin. The Swedish-Russian financier had been pondering the possibilities of the new oil-burning engine as it related to his Russian oil holdings and the potentials of manufacturing Diesel engines for sale and use in Russia.

Thus, on February 16, 1898, still another international franchise came into being. By its terms Diesel received 600,000 marks in cash and 200,000 marks' worth of stock in the newly founded Russian Diesel Engine Company of Nürnberg, in return for granting Emanuel Nobel and associates the right to manufacture Diesel engines and sell them in all the Russias.

After his successful meetings with Nobel, the inventor traveled to England to negotiate a licensing agreement with the firearms and munitions maker Sir Hiram Maxim, who was regarded as the "cannon king" of Britain, much as Krupp was of Germany.

That journey brought pertinent testimony of the status quo of the Diesel headaches. In a letter to his wife in Munich, Rudolf said that he had slept the previous night without the customary sleeping powder. By then he was taking large doses of antipyrin, but its efficacy appeared to be only temporary. As what is called success crowded upon the inventor, his health problems grew more trying. As his life grew opulent, his headaches became persistent and more severe. The exact cause was still unknown. There was, of course, the factor of eyestrain. Since his early thirties, Diesel had worn glasses—rimless "pinchets"—almost continuously. Yet his nagging headaches kept getting worse, and medical counsel seemed of little avail.

From his side-street apartment in Munich, poor old Papa Theodor continued to worry his rich and famous son with suggestions for magnetism treatments. If Diesel followed the suggestions, which seems doubtful, he was not helped by those treatments either. Between worries over business and health, Diesel was losing much of his tranquil charm.

Diesel judged that the Scottish license holders had done disgustingly little about marketing his engine. When he suggested that the Scots were dragging their feet, the directors of the company replied that the Diesel engine in its present form was not quite as dependable as it should be. Accordingly, the Mirrlees firm was moving along slowly with manufacturing. Impatiently, Diesel offered to purchase their licensing agreement from them for fifty thousand marks more than they had paid him for it. The cautious Scots refused. Although the Mirrlees firm had completed a model engine which ran quite well, several minor problems had developed, any one of which could prove an irritant to a customer, certainly a Scottish customer. As other foreign manufacturers had discovered, there were no mechanics or engineers who understood the engine nearly as well as Diesel or nearly well enough.

Resignedly, the inventor returned to Munich and busied himself with manufacturing problems related to his engines in the various licensed plants of Europe. Constantly on the move from one manufacturer with his particular problems to the next, Diesel showed symptoms of exhaustion. Astutely he trained a few men in the care and repair of Diesel engines. Karl Dietrich assisted in the installations of the French and German Deutz engines; Ludwig Noé worked with the first Swedish, Danish, and Hungarian machines; Hans Erney, with those in Britain and Belgium; and Anton Böttcher, with the Swiss and American Diesel engine installations. All were "Diesel trained" and employed.

The "first-aiders" helped reduce the inventor's worry load, or at least one part of it. But there were other parts. The patent controversy threatened earlier by Emil Capitaine began to flare once more. Diesel's nerves were so very much on edge that his doctors ordered a complete rest, having apparently agreed on a diagnosis of *neurasthenia cerebralis,* or nervous exhaustion. The patent fight with Capitaine went on in the press, on lecture platforms, and in various technical journals.

The *Reichspatentamt* heard the case and on April 21, 1898,

turned down the appeal of Emil Capitaine on the grounds that it could find no actual infringement of any previously existing patent.

But Capitaine intended to take his case to the higher authorities. More speeches and pamphlets attacking Diesel followed. It seemed that this protracted litigation would take up much time and consume tremendous energy. In July, Capitaine demanded twenty thousand marks from Diesel, in return for which he offered to cease further prosecution of his case. The inventor turned down the demand. Until his death in 1907, Emil Capitaine remained Diesel's bitter enemy and relentless antagonist.

Fortunately there were lighter moments. Many of them were associated with Munich's own Second Power and Works Machine Exhibition which opened its wrought-iron gates on the first day of summer, 1898.

Rudolf Diesel took his wife and their three children to the fair. It was, as the slender, easy-spoken, worry- and headache-ridden inventor reflected, high time he was renewing acquaintance with his own flesh and blood. Daughter Hedy, pretty, winsome, and pleasantly curvaceous, resembled her father at least as far as facial features. Sons Rudolf and Eugen were both growing into handsome boys, slender and erect like their father, but both with features more like their mother's. Both, like their father, were meditative.

A photograph made at or near the time of the Munich exposition, shows the five walking in a somewhat windblown rank, the father and daughter abreast, the mother and two sons following about a half-pace in the rear. Martha Flasche Diesel had the appearance of being a wifelike Madam Moonlight who seemed never to grow old and only slightly plumper. Rudolf appeared to be a proud but somewhat self-conscious father grimly determined to be carefree however much it might hurt. Both the sons seemed to register a rather baffling quality of sadness. "Little" Rudolf, who by then was shoulder tall to his six-foot-one father, looked quizzically ahead with an expression of resigned doubt. Eugen seemed reflective and suspicious of the photographer's intentions.

Munich's exposition brought forth a bumper crop of cameras and camera addicts along with festive tourists and serious hawkers. Although by no means the biggest of what had materialized as an international epidemic (the medical term is pandemic) of expositions, Munich's trade fair was unquestionably the most extensively photographed and probably had the best food of all.

One of its widely heralded innovations was an "automatic photo-machine." When one deposited a coin and faced the magic box, there was a bright flash of light. The resulting picture was imprinted in a small metal plate, which a few minutes later dropped out of the "bath" in the under box. Apparently all the Diesels tried the photographic miracle. The father was both an ardent photography fan and an able photographer.

There was a great deal to behold with wonder, pleasure, giggles, or squeals or any combination thereof. A great and gay midway spread over a well-graded peninsula which tongued far out into the Isar River.

Its most "spectacular" feature was a sloping water chute down which a wheeled wagon or lorry rolled at breathtaking speed. Customers dressed in proper bathing attire of the times, which was fully as all-covering as the prevailing evening dress, crowded into the lorry and braced themselves for the "thrill of a lifetime." Faster and faster the vehicle would rocket downward—until it plunged into the river with a tremendous splash. As passengers gasped or shouted or squealed their reaction to the high-speed dunking, a power winch dropped forward, latched onto the lorry, and pulled the conveyance back up the ramp and so back to the changing room. The oversize and accelerated ducking stool was then set for a repeat performance.

For those opposed to high-velocity dunking, including the male Diesels, there were other spectacles.

Beginning at dusk a giant fountain threw intricately patterned sprays of water high into the air duly illuminated with vari-colored spotlights. Over the teeming fairgrounds floated the sausage-like balloon of Major Parseval, who was later to build a first rigid airship.

Several automobiles built by Daimler and Benz were also exhibited and at given hours were driven, sputtering and fuming, about the spacious and well-groomed grounds, to the mingled delight and disdain of human onlookers and the consternation of the horses, which seemed to sense and resent the undermining of their age-old domain.

As the name "Second Power and Works Machine Exhibition" indicated, machinery comprised the more significant exhibits. For the first time in Europe Diesel engines were on public display, and in a truly big and impressive way.

Diesel led his family proudly to the front of the special pavilion which housed the *Kollektiv-Ausstellung von Diesel-Motoren.* He pointed out the tall exhaust pipe on the roof of the building, out of which every second came a burst of barely visible exhaust. There was no heavy black smoke such as the steam engines made.

Inside the specially built, light-brown wooden pavilion stood four mighty Diesel engines, three of them running with a steady, power-throbbing rhythm. The *Maschinenfabrik* Augsburg was also showing a 30-horsepower Diesel engine which operated a pump patented by Brakeman. The Fried. Krupp *Werke* of Essen exhibited a Diesel engine which developed thirty-five horsepower to turn a high-pressure centrifugal pump made by the *Gebrüder* Sulzer *Werke.* Both pumps sent great streams of water cascading into the nearby river.

The *Maschinenbau-Aktiengesellschaft* Nürnberg was represented by a 20-horsepower Diesel engine set up to demonstrate the starting, regulating, and smooth-operating qualities of the power producer, as well as its braking and ignition competence. *Gasmotorenfabrik* Deutz of Cologne exhibited a Diesel engine which also developed twenty horsepower and operated a Linde air-liquifying machine.

The liquid-air machine attracted particular attention. Even though no use was then known for that most unusual product, visitors found the evidence that invisible air can be changed to liquid edifying, even startling. As the astonished spectators gaped, an attendant dipped a piece of flexible rubber tubing into the cold liquid and showed how the tubing almost instantly froze so hard that it could be splintered by a hammer blow.

All four of the Diesel engines were single-cylinder "power plants." A twin-cylinder engine, developing forty horsepower and equipped with a Schuckert dynamo, was to have been displayed, but the Nürnberg factory was unable to complete it in time for the exposition.

All the Diesel engines on exhibit had been assembled at their respective manufacturing plants in great haste and had to be regulated after installation at the fair. As one might expect, the cold engines at the exhibition did not always perform according to schedule or expectations. Since the newfangled power builders were inherently difficult to start, an engineer would arrive on the scene each morning before the gates were opened to the public, start the obstinate engines, run them until they were well warmed, and then shut them off. When the exhibition hall was opened to the public, the engines would ordinarily, though not invariably, start fairly easily. This and the no less remarkable fact that the engines usually kept running were a tribute both to the practical merits of the machines and to the skill of the engineers who had assembled them in such a very short time.

At that, the exhibit had its tense moments. The fair's most eminent visitor was Prince Ludwig, later King Ludwig III of Bavaria. The Prince was eager to see these "constant-pressure engines" in action. His Royal Highness—*Seine Königliche Hoheit*—had been told that they ran with considerably less noise than the more familiar combustion engines. Unfortunately something went wrong that day; not one of the exhibited Diesel engines was operating. It may have been the fuel, or, to quote a fine old mechanic's diagnosis, the engines may have been "behext." The Prince came into the pavilion, inspected the silent, non-operating engines at considerable length and in deep thought. Then His Royal Highness commented that these Diesel engines were indeed quiet; one might say, *deadly* quiet!

After the exhibition closed that fall (1898), work was resumed on the basic or pilot Diesel engine in Augsburg. The many technical inspections and detailed examinations which repeatedly resulted in complete dismantling and reassembling of the engine made further systematic tests of individual phases or improvements of parts practically impossible. By then the Diesel-trained mechanics were either worn out by engine tending at the fair or were assembling engines to sell. But the inventor kept stubbornly gathering data about the characteristic behavior of the engine.

In due time he saw his "old engine" restored to operational status; its cylinder was rebored, all parts scrutinized and renewed as needed, and a new type of starting valve was made and installed, as was an adjustable connecting rod for changing the compression ratio. Then Diesel saw the engine, designated as "A-Motor," made ready to be used exclusively for experiments with gas fuel. The other "early engine," which had been built by the Augsburg works and shipped to Krupp as a model, was returned to serve as "B-Motor"—for experimenting with liquid fuels. Later, along with many valuable papers and diagrams, drawings and charts, "Motor B" was presented to the Deutsches Museum at Munich. Unfortunately earlier engines and test pieces of historical interest were not saved and cannot be found by engine researchers.

As the year hurried along, Diesel decided it was indeed time for him to form a company for developing foreign manufacturing rights to his engine. By 1898, the work involved in extending rights was far too strenuous for Diesel to carry on alone. With the assistance of a group of financiers, the *Allgemeine Gesellschaft für Dieselmotoren, Aktiengesellschaft,* was founded on September 17 to "internationalize" the Diesel engine. The corporation would also undertake further development and improvement of the Diesel engine and therefore required all plans and papers by the "construction bureau." It was agreed that as compensation for basic patents and other rights Diesel was to receive 3,500,000 marks in cash at the time of formal registration of the incorporation papers. Actually he received less than half that amount, only 1,250,000 marks in cash and the rest in stock certificates of the newly formed enterprise.

At least on paper Rudolf Diesel was a millionaire five times over by his fortieth birthday. By any rational or reasonably rational appraisal the dreamer with the poet's face, the mathematician's mind, and the oil-smeared mechanic's hands had gone far. He was determined to go much farther.

Theory Versus Workability

THE GENERAL SOCIETY FOR DIESEL ENGINES (*Allgemeine Gesellschaft für Dieselmotoren*) took over the crucial task of gathering data on the operating Diesel engine. Karl Dietrich took charge of the test work and record-keeping. Immanuel Lauster, representing the Augsburg factory, took care of the necessary shop work. Both men were assisted by Max Ensslein and Karl Grosser. Later Heinrich Philipp replaced Ensslein, and Paul Meyer took over from Dietrich the direction of the work. Others actively engaged in the research and development included Fritz Reichenbach, Anton Böttcher, Rudolf Pawlikowski, Schüler, Lietzenmeyer, and Ludwig Noé.

During his Paris days, Diesel as a "cold-engine man" had experimented extensively and familiarized himself with characteristics of many kinds of crude oil from many countries. At that time he had been especially interested in extracting paraffin from petroleum by applying extremely low temperatures. As already noted, Diesel had designed the first engine for fueling by liquid oil; in the experiments of five years before (1893) the thick *Pechelbrenner* (crude oil) had first been used. Early tests with the Diesel engine had included the successful use of gasoline and both Russian- and American-made lamp kerosenes. But when difficult problems arose, the inventor decided, and wisely so, to get on with perfecting a workable engine, then to return to the problem of selecting or developing versatile and economical fuels.

During 1897, Diesel and his colleagues had returned to experimenting with other fuels. By then they were able to test in terms of continuous operations of the engine and accurately diagram, chart, and measure the results: ignition reactions, combustion in terms of the atomization process and spraying mechanism, texture and odor of exhaust gases, the cleansing effects of carbon deposit on engine parts, and so on. They had begun testing heat values, water mixtures, chemical additives, burning characteristics, and the like in the open air. They were also checking the sources, availability, and prices of all possible fuels.

At first it had seemed that each different liquid fuel required a different injection mechanism and jet apparatus to insure efficiency. The experimenters mixed some of the heavy oils with the thinner and easy-flowing oils. Although gasolines, refined benzine and ligroine, had already worked well in the engine, all were expensive. However efficient, none was sufficiently economical for extensive use. In the late 1890's, heavy kerosene or lamp oils carried a high import duty (15 marks for 100 kilograms—approximately $1.75 a hundredweight, or 15-plus cents a gallon). For German and other European use, this was virtually prohibitive.

In terms of both price and availability, waste oils—the so-called red oil, yellow oil, dark paraffin oil, and solar oil, all inexpensive by-products of the flourishing domestic soft-coal industry—were definitely desirable; but the high viscosity of these oils presented many vexing problems. The spray mechanism clogged easily, carbon deposits were heavy, and the engine would not start unless primed with a more readily ignitable fuel. Eventually these problems were solved, and the heavy soft-coal oils were proved to be suitable fuel. Measured by the horsepower hour, the improved Diesel engine consumed only ten grams (one-third of an ounce) more of the soft-coal extract than it did kerosene.

Somewhat ironically, in America, where kerosene was becoming comparatively cheap and plentiful, first tests showed better results from the medium soft-coal oils than from refinery-graded kerosene. In Germany, various Russian solar oils, Galician blue oils, and *Pechelbrenner* solar oils were being tested and found to be suitable for use in the Diesel engine. The "real raw" oils, such as raw naphtha from Baku, raw well oil from Romania and Galicia, as well as the German petroleum from Tegernsee and Oelheim in Hannover, also proved satisfactory for use in the engine.

Greater difficulties were experienced with certain naphtha waste products, such as the so-called Russian masut from Baku, which was acquired through Nobel. In Russia this refinery sludge was being used widely as fuel. Ignition and combustion were passably good, but its viscosity or lack of ready flow required that it be mixed with a thinner oil.

Experiments using pure alcohol fuel were also begun. Presently the experimenters learned that starting the engine with kerosene, then changing the running machine to alcohol, with due changes in the procedures of fuel injection and air density, worked out reasonably well, though alcohol fueling was still too expensive for extensive use.

Hard-coal tar oils, creosote and benzol, in their original form or mixed with other oils, were also tried, but were especially difficult to work with since their composition, often erroneously stated by their suppliers, was far from stable. Every barrel seemed to present a different problem. Not until later was it discovered that hard coal and its oil by-products are decisively affected both by variation of temperature and by the form and position of the original geological deposit.

A special situation arose in France, where the exceptionally high tariffs levied by state and city made imported fuels prohibitively expensive and forced the use of reasonably priced fuel producible within home boundaries. Under the guidance of Frédéric Dyckhoff, the Diesel Engine Works at Bar-le-Duc concentrated on testing tar oils exclusively and eventually proved them to be satisfactory fuels.

Since no interest in the use of shale oil was shown in Germany, this product was not tested. However, in England a Scottish-developed shale oil was used successfully in the first Diesel engine operated there. Stationary engines in France were also fueled with the same type of oil, which was easily procured and was inexpensive. This *huile de schiste* was reasonably priced and functioned well.

As the nineteenth century ended, it was obvious that the fate and scope of the internal-combustion engine were dependent on its fuel or fuels. At the Paris exposition of 1900, a Diesel engine, built by the French Otto Company, ran wholly on peanut oil. Apparently none of the onlookers was aware of this. The engine built especially for that type of fuel operated exactly like those powered by other oils.

In 1894, when Diesel failed to achieve the desired results with liquid fuel, he was experimenting with a combustion chamber located within the piston in which illuminating gas worked best. The results with illuminating gas were good in a later engine in which the combustion chamber was located in the cylinder head where air was introduced to burn the fuel. But now that the combustion chamber was located within the cylinder itself, illuminating gas seemed to lose its effectiveness.

Therefore, results with illuminating gas were not as good as those achieved five years before. Strides had been made in developing other gas-burning engines whose thermal efficiency matched or almost matched that of the Diesel engine. By the end of the 1890's, it was quite evident that the Diesel principle, even if successfully adapted to gaseous fuel, could not muster enough advantages over other less costly gas-burning engines. As a result, Diesel men turned finally to liquid fuels.

Before concluding the entire fuel-testing program, however, they made several noteworthy experiments with coal-dust fuel.

In his speech at Kassel in 1897, Diesel had suggested that his engine would reach its "fullest importance" only when common hard coal could be used as its fuel. During the seven-year period after the idea of using coal-dust engine fuel had sprouted, a number of notable developments with coal fuels had emerged. The use of coal dust for stoking boilers had been proposed with interest and enthusiasm. But practically speaking, no extensive coal-dust industry had yet developed. Coal dust for use in various new and highly promoted "special stokers" was not nearly fine enough for use in any combustion engine visualized by Diesel. To be effective in any of his test engines, the dust had to be as fine as the finest cake flour and capable of remaining suspended in still air for a considerable period of time. To grind and sift such extraordinarily fine coal dust would be too expensive and complicated for practical use.

Even worse, because of its hygroscopic character, the coal dust's affinity for moisture caused it to coagulate, thus losing heat value and burning properties. Storing the fine coal powder also involved a rather serious fire or explosion hazard.

Finally, as Diesel was well aware, to adapt powdered coal for use as a staple fuel would involve expensive redesigning and virtual rebuilding of his engine. There would have to be a new type of injector mechanism, different lubrication for the piston, restoring of the combustion chamber outside the working cylinder, and remodeling and rearranging most other operative parts.

Any conventional engineer would have been entirely willing, indeed eager, to drop the complexities of powdered coal fuel and follow the way already open to the practical use of fuel oils. But Diesel was not a conventional engineer. He believed in *toto* the text of his address at Kassel. He therefore set out to win the directors' approval of more research with powdered coal fuel. Once more the Diesel persuasiveness won out—with much more hard, exacting, and expensive research the result.

By 1899, Diesel had approval to begin experiments with coal dust. As a first step, the powdered coal was mixed with indrawn air and compressed. The "first reduction" was then enriched with a small quantity of igniting fuel and blown into the cylinder through the already established atomizing mechanism at top dead center. In time the mixture ignited and burned quite well.

This success was achieved, but not without a great many exacting preliminaries. First, the ignition properties of various types of coal dust had to be tested in open air. Then precisely measured quantities were blown by means of a blow tube into a glass cylinder. Suspension and settlement of the dust were carefully noted. Commonly available coal dust was sieved until extremely fine and then blown into an open flame, where it exploded before burning almost clean. Diesel and his helpers next constructed an umbrella device which they placed above the flame to catch the residues of very tiny coke particles. This proved that only the extremely small particles burned completely. The principal portion left substantial quantities of coke as residue.

For the tests within the operative engine, an apparatus was acquired from the Swiss Coal Dust Firing Corporation (*Schweizerischen Kohlstaubfeurungs-Aktiengesellschaft*) at Zurich. The dust was poured into a funnel on the base of which was a vibrating sieve of 494 mesh a square centimeter (0.155 a square inch). The Diesel engine was started with kerosene, then switched to the coal dust. It ran for five minutes. Then a takedown and examination showed the piston covered with a thick coating of unburned coal dust. After the next test run of seven minutes the inside of the valves was burned completely clean. The combustion chamber and the piston crown were covered with only a very light soot, closely comparable to that left by kerosene fuel. Even so, the preliminary experiments showed that rather extensive redesigning would be required to develop an effective coal-dust burner.

At that point Diesel made another rather astonishing move. Instead of pushing on at the stubborn and expensive procedure with a reasonably good chance of developing still another enormously valuable property from his basic patent, he permitted one of his engineer assistants, Rudolf Pawlikowski, to leave his employ (in February, 1898) to form his own company, *Kosmos G.m.b.H.* (*Gesellschaft mit beschränkter Haftung*, a company with limited financial responsibility; equivalent to Ltd.), at Görlitz to continue work on and perfection of a practical coal-dust engine. Pawlikowski eventually developed such a Diesel engine.

Although this other Rudolf never completely solved all the

tough and intricate problems of fueling with powdered coal, he did complete the valiant effort, even though it fell short of being commercially successful. One gathers that if Diesel had not been obligated to and dependent on others, he would have pursued the experiment with powdered coal himself, even to spending his last mark. But Rudolf Diesel centered his tremendous energies on another goal. He already held a first patent, dating back to February 28, 1892, on a compound engine. After eight years and, one deduces, a vast amount of study, the inventor sent to the *Reichspatentamt* drawings for a proposed engine of about one hundred horsepower.

It consisted of two working cylinders, identical to those of a one-cylinder engine but connected by carefully timed valves with a larger center cylinder. By means of other timed valves all the cylinders were connected with a large air tank. The piston in the larger, or center, cylinder sucked atmospheric air through a valve from the outside and during the upstroke compressed the "free air" and forced it into the tank. The lower part of the center cylinder operated as an air pump and created the pre-compression of the combustion air. Water sprayers were located near both the inlet and the outlet air valves so that the operation could include water vapor, if desired.

The operation of the working cylinder was generally similar to that of the earlier single-cylinder engine. But there was one important addition: the piston sucked the pre-compressed air out of the air tank instead of out of the atmosphere. On the upstroke the piston accomplished the second stage of compressing the air to its prescribed pressure. Gradual fuel injection took place at the downstroke. At the lowest point of piston descent, the valves opened as the piston of the center cylinder was at its highest point. When the piston in the working cylinder went down, the piston of the center cylinder went up and the combustion gas expanded to fill the center cylinder to capacity. With that, the valves of the working cylinder closed, and the top center valve opened, so that with the next upstroke of the piston the combustion gas "exhausted" into the atmosphere. The basic goal of the compound engine was better utilization of heat than was possible with only one cylinder.

First construction drawings for this compound engine had been completed by Johannes Nadrowski in Berlin in 1895, and later revised in Augsburg to take advantage of various lessons learned from testing the one-cylinder engine. Since the lengthy tests of the already completed one-cylinder engine made formidable demands on the rather small work force then available, Diesel did not try to rush the completion of the compound engine. Instead, he worked along toward the rather deliberate completion of the first pilot models. This required an entire year. Another six months passed before the first model of the compound engine was installed in the Augsburg laboratory near the improved single-cylinder engine.

The compound engine was, comparatively speaking, a giant. The diameter of its working cylinder was 200 millimeters (7.87 inches) and that of the expansion cylinder, 510 millimeters (20.08 inches). A piston stroke of 400 millimeters (15.75 inches) was produced by the piston connecting rod 80 millimeters (3.15 inches) long. The engine operated at 150 revolutions per minute. At first it was found that the vessel air cooled too much between low and high pressures to permit dependable operation. To eliminate this fault, the air was channeled directly from one cylinder to the other. But with this alteration the high-pressure cylinder overheated, and the cooling water boiled away. Consequently the hot piston growled loudly, and the transfer valves warped.

With these troubles corrected by means of revised diameters and compression ratios, the engine ran for several hours without serious disturbance. However, tests indicated a considerable pressure loss during the transfer of the burned gases from the high-pressure cylinder to the low. To recover this loss, the inventor and his helpers devised means for retaining a part of the exhaust gases in the low-pressure cylinder.

This, however, interfered with effective compression. A formidable heat loss (of 11.8 per cent) was being caused by the transference from one cylinder to the other. Diesel had conceived his dual engine principle with the help and guidance of pure mathematics, but the turbulence of air and many other gases knows few absolute rules.

Even more than the project to perfect an engine which would burn powdered coal, the compound engine appeared to be a magnificent undertaking. But it did not work out. Theoretically, it was sound, as most of the engineers agreed. But the gap between theory and practical work could not be bridged. Diesel grew painfully aware of this fact. He was no less painfully aware of other dire developments.

His headaches were growing steadily worse. By 1898, he was making mention of his failing health. His actual work records did not substantiate the reference. In no year of his busy life had he worked harder. The difficulties connected with his proposed compound engine seemed to add to his certainty that his first and basic engine design would continue to prove itself a wise investment.

Others shared his conviction. The Augsburg-Krupp syndicate had spent the precisely recorded sum of 443,335.31 marks (over $105,000) for building, developing, and testing the Diesel engine. For the four years 1893 to 1897, Diesel received 30,000 marks yearly from the combine. Other incidental expenses incurred brought the total expenditure to approximately 600,000 marks. Most of these expenses were shortly recovered through licensing other manufacturers in Germany. But the fact stood that Rudolf Diesel had signed his invention away for a price. For the manufacturers the price was certainly not too high, but the money received was not nearly sufficient for the catapulting needs of Rudolf Diesel and his now appallingly expensive family and home life.

The Uncertainly Rich Man

WORKING DOGGEDLY, even fiercely, in the Augsburg plant and laboratory, Diesel continued to espouse fair treatment for factory labor and tried to put what he termed the "human honor system" into effect, with far from comforting results. Foremen charged him with excessive leniency —pampering the hired help. Although he steadfastly denied the charge, he did, however, begin to notice with increasing resentment that workmen, including assistant engineers, were taking advantage of the laxity of discipline. The gentle inventor then tried firmness. He heard himself shouting strong words—an act which he detested: "I thunder at them like a *Donnerwetter.*"

This statement was probably an exaggeration, but Diesel insisted that his headaches were beyond possibility of exaggeration. They nagged and tortured him, accompanied him on his incessant business trips, and added torment to his patent litigations and his interesting but frustrating experiments.

His wife and his children, meanwhile, seemed to be growing away from him and he from them. Martha greatly enjoyed the social advantages which her husband's newly gained money and prestige made possible. She deplored quite understandably his frequent absences from home and family. She remained rather tolerant toward his engine—his "black mistress"—which so greedily consumed his time and interest. But she appeared to be losing her enthusiasm for the engine and his career. Yet, by her own standards, Martha Diesel remained a good wife.

How good a wife for Rudolf Diesel was another question. By the late autumn of 1898, the forty-year-old inventor was undeniably a sick man. His pain was excruciating, yet extremely difficult to diagnose. As an intensive researcher, Rudolf Diesel continued to find many facts, but he did not find inner peace. His search for it was almost constant but futile. Martha did not seem to be able to comprehend this search.

During October, Diesel's doctors strongly advised him to enter a private sanitarium in Neuwittelsbach, near Munich for an extended period of complete rest. Reluctantly but resignedly Diesel agreed. It may have been a bad decision. For one accustomed to an active life, the abrupt change to meditative indolence was not necessarily wise. As events were to prove, there was peril from a reversion to the shadowy mysticism which had engulfed his father and his grandfather.

Diesel took with him designs for a magnificent house he and Martha were planning to build on Maria-Theresia-Strasse in Bogenhausen, a fashionable suburb of Munich. Shortly before Christmas, he was permitted to return home for the holidays. Apparently they were not especially happy. Certainly the headaches persisted. Soon after New Year's Day, Diesel returned briefly to the *Heilanstalt* (sanitarium). Later in January (1899), still suffering from unbearable headaches, he agreed on doctor's advice to take another and longer "rest cure," at least three months' recuperation.

This attempt likewise produced no cure, but the setting was diverting. The chosen spot was Meran, capital of Tyrol, about thirty-six miles south of the Brenner Pass. The ancient town is located in a high valley around which tower the usually white-capped Dolomites with peaks ranging from one mile to two miles above sea level.

In somewhat mixed Teutonic metaphor the town and its environs have been described as a "High Pocketful of Dreams." The recorded history of Meran dates back to 1300; that of its magnificent romanesque entrance gate, Vinschgauer Tor, even farther. There are splendid mountains surrounding the ancient highland capital. One after another, the lower slopes were and, to a less extent, still are dotted with castles with engaging names blending Latin and ancient Saxon—Fragsburg, Zenoberg, Auer, Brunnenburg, Schloss Forst, Plonta, Schenna, Rubein. Diesel found the ancient castles and peoples as captivating as the magnificent landscape.

Following his train journey to Meran, he viewed the Castle of Zenoberg, which had withstood the occupation of the Roman legions. He saw stone bridges and one-time city walls which had been standing for perhaps a score of centuries. On the little green square directly in front of the railroad depot stood a statue of the great Tyrolean liberator Andreas Hofer, the martyr from St. Leonhard.

Julius Caesar had looked down on and ably described in his *Commentaries* the same undying dream world. Warring conqueror or not, this slender and tense Caius Julius had been an all-time great internationalist, and the Meran country is a timeless haven for internationalism. Although now part of Italy, its people remain, first and last, Tyroleans, with an ineradicable feeling for Austria and a poetic interest in holding an unchanging view of a changing world.

Diesel seemed to absorb the mood of the ethereal community which reaches back so many centuries. He took quarters for two dream-splashed months in Schloss Labers. The castle overlooked the ancient town and the great sweep of the Etsch (or Adige) Valley, into which the Passer (Passirio) River flows from the north. Diesel's liking for the Shangri-La was "in character." He was and throughout his life had been very much an artist. He loved the beautiful.

Although blessed with magnificent scenery, the castle provided lodging characterized by a pompousness which only a few years earlier the inventor would have discreetly avoided. As a "mansion inn" the castle reeked with uncomfortable sumptuousness. The rooms were huge, ceilinged like superannuated opera houses and furnished with grandiose and antiquated magnificence. Fellow guests included a financial tycoon, a judge, a

Prussian army officer, and the inevitable waxen bouquet of elderly ladies of means with nothing to do.

The time was winter, for which the castle's heating system was deplorably inadequate. Along with the other wealthy boarders, Rudolf Diesel shivered and sneezed until April came. Then he returned to Munich and to work.

He began May by setting out on a trip to Austria-Hungary, where meeting followed meeting with almost sadistic urgency. In a typical, brief letter to his wife, the inventor jotted: "Thursday, ten o'clock, meeting of our *Dieselmotorengesellschaft*. Lunch, tour of a second engine factory. From there to a meeting at six o'clock on petroleum matters until seven-thirty . . . . Walk to the hotel to change clothes, until eleven-thirty a fine dinner, to bed at one o'clock, after seventeen hours of meetings without interruption with two usual meals. It is miserable: Slept three hours. Friday, rose at five o'clock. At six rode to Arad, through the Hungarian *pusta*. Arrived at twelve o'clock, celebrity reception by a local engine manufacturer, drive into town, usual dinner, tour of the plant facilities. Four o'clock return trip to Pest, where finally my good friends left me. Terrible heat."

His wife answered: "Your messages have made me quite dizzy; what you had to achieve is unheard of. I have read your letter very attentively. How will you come home to me! To stand by, not to be able to hinder or to help, that is my fate. And the hope that the petroleum affair is your last one, I have given up. A man like you pushes on *(wühlt weiter)*."

"*Wühlt weiter*-ing" (or, in correct German, *weiter wühlen*) involved a great deal besides the "petroleum affair"—an ill-fated investment in a Romanian petroleum development. Some six months later Diesel wrote to his wife from Vienna that "in order to expedite the petroleum affair," he would go directly to Lemberg (Lvov), and from there "tour the actual oil fields to see for myself how the situation stands." He went, saw, and found that the situation was not standing. His investment in oil stock, a speculation which may have been incubated during the course of his stay at Schloss Labers, turned out to be a bone-jolting loss of perhaps one-third of a million marks.

The swift, hard blow was followed by others, less swift but no less damaging. As a commercial product, the Diesel engine was having a difficult time getting started. The inventor was finding himself more and more handicapped in taking direct or effective action, for by the end of the century he had lost proprietorship of his great invention along with lesser ones. The future of the Diesel engine was now in the hands of the directors of the *Allgemeine Gesellschaft für Dieselmotoren* and, less directly, of its stockholders, of whom R. Diesel was but a minority of one. The inventor had no direct say about the corporation's plans for the future. He had only the minor stockholder's rather dubious consolation of knowing that the "company" was, or appeared to be, doing passably well for itself. Sales of Diesel-engine manufacturing rights for Austria-Hungary alone had lately netted 1,250,000 marks. By coincidence less than happy for the inventor, that sum was the exact total which he had received from his "company" for his most historic invention.

But in an appalling number of instances, the earth-changing engine was only changing bank balances from black to red. The unhappy process was being demonstrated on home grounds— at Augsburg.

Up to this time the new *Dieselmotorenfabrik* had built only a few engines. One of which had been sold to a brewery which had promptly returned it; it just didn't run right. The factory was seemingly not managed well or staffed competently. Soon other engines which had given unsatisfactory service were being heaped in the factory's receiving yard. The directors of the Augsburg company failed to take prompt or positive steps about the situation. Rumors spread, as rumors will, that all or practically all Diesel engines were unsatisfactory. The stock of the corporation, which had climbed rapidly, now tumbled abruptly.

Diesel proceeded to the factory and consulted with engineer Ludwig Noé, who was an expert on the engines. With the inventor's help Noé built a model which ran quite well. But it was too late to save the *Dieselmotorenfabrik* from collapse. As a frantic last hope, the firm sought to put into production a two-stroke-cycle Diesel engine developed by Hugo Güldner. As is usually the case, the desperation move did not succeed. The two-cycle engine failed to function properly, so that further testing and experimentation were required. The factory had planned to manufacture a speed pump powered by a Diesel engine, but the production of this unit never materialized. The Diesel engine factory at Augsburg collapsed and died.

Late in 1898, the pioneer *Maschinenfabrik* Augsburg had merged with the *Maschinenbaugesellschaft* Nürnberg, A.G., to become known as M.A.N. (*Maschinenfabrik* Augsburg-Nürnberg). By the following year this company also experienced difficulties with some of its first-delivered engines. Many more of the early machines developed minor operating difficulties, and often an experienced engineer had to be sent to correct an error in construction or replace a malfunctioning part.

Immanuel Lauster, who had taken over the direction of Diesel engine construction at the M.A.N. factory, continued making minor changes and improvements, thereby serving the total cause by making engines which performed consistently and reliably. Lauster, a pioneer champion of the Diesel engine, continued to be one of its most able perfecters.

There were not enough Lausters to go around, however. The Deutz works, for example, had built a series of thirteen Diesel engines but did not deliver any of them to customers. The first engines did not operate to the maker's entire satisfaction, and the company decided to test them extensively before putting them on the market.

In St. Louis, where beerman Busch continued to reign grandly and prosper, the Busch-financed American Diesel Company was having formidable difficulties keeping up with the many rapid improvements being made in Germany. The surviving Augsburg factory was setting an example of enlightened patience. M.A.N. steadfastly refused to sell any engine until it had been proved by almost innumerable tests over a period of at least six months and could therefore be labeled practically foolproof. The Busch company was less patient and more set on rapid improvements. The attitude was honorable, but the practice was not successful, even though expensive and laborious.

In spite of reports of successful engine production from Dyckhoff in France and Nobel in St. Peterburg, Diesel, understandably, was deeply concerned with homeland failures. Maintaining more than usual calm, he traveled across Europe, talking to the perturbed manufacturers, counseling them, bolstering flagging enthusiasm, and urging them not to give up because of temporary setbacks but to hold to their belief in the machine's ultimate success.

It was good thinking and good talking, but it did not produce immediate benefits. The hoped-for exchange of information concerning technological advances and improvements made on the engine by the individual license holders did not materialize. Diesel had hoped that the builders would help each other by an exchange of new ideas and on-the-job discoveries. This cooperative spirit was a reflection of his own nature, not the licensees'. The competitive feeling among the various manufacturers was too great to encourage or even permit mutual assistance. But the inventor kept hoping, working, and traveling.

The benefits, if any, of his rest at Schloss Labers were quickly fading. Progressively the inventor took more and more medicines in the hope of soothing his frayed nerves, curbing his headaches, and inducing sleep.

But now there was more than manufacturers, franchise jumpers, patent infringers, and broken-down engines to worry about. Another world's fair was in the offing. Another "spectacular" Paris exposition was set to open on April 14, 1900. Its special significance was Europe's awareness of the fast-rising importance of the New World across the Atlantic. The United States was second only to France in number of exhibits at the exposition—a grand total of 6,564. As the new century dawned, the industrial productivity of the United States was more than twice that of

Great Britain, which had been helmsman of the Industrial Revolution.

The "greatest" Paris fair opened with impressive overtures. Its grounds were big, but from the opening, its crowds were even bigger. Very early in the season an approach bridge collapsed from being overcrowded. A great-grandmother, more than one hundred years old, from Seganne walked for fifteen days to attend the fair, thereby setting a duly publicized precedent for attendance, which went over the fifty million mark.

As the frantically busy Diesel was well aware, the impending Paris exposition, while hailing the new century which he knew would encompass unrivaled wonders, was also a carnival closing for a century of monumental inventions and developments.

Almost all of the latter had emerged during Rudolf's own lifetime; the rest during his father's lifetime. Theodor Diesel was twelve years old in 1842 when the American surgeon Crawford Williamson Long first used ether as an anesthetic. Rudolf was the same age when England's great surgeon, Joseph Lister, introduced antiseptic surgery. He was twenty-four when Louis Pasteur discovered a treatment for rabies and when Robert Koch isolated the micro-organism of tuberculosis. By 1894 another generation of Diesels was of age to benefit from Paul Émile Roux's near perfection of an antitoxin for diphtheria.

Throughout the literate or quasi-literate world, men of science and non-science were still warmly opinionated on the principle of evolution and the theory of natural selection advanced by England's Charles Robert Darwin as early as 1859. The "laws" of heredity were hypothesized by Austria's botanist Gregor Johann Mendel just as the appalling American Civil War was closing but were little noticed until about nineteen years later.

In the course of 1899, a year all but delirious for R. Diesel, the discovery of radium by the French chemist and physicist Pierre Curie and his wife, Marie, was quite properly the most important scientific headline. By contrast, Frenchman Antoine Henri Becquerel's discovery of the radioactive nature of uranium three years earlier had been noticed by only a few students.

As a thermal engineer, Diesel was keenly interested in his century's milestones of applied chemistry. These included the grand first steps, such as American Charles Goodyear's invention of a process for vulcanizing rubber in 1839 and, ten years later, the German chemist Christian Friedrich Schönbein's invention of guncotton. England's Henry Bessemer's development of the air-blast converter for the manufacturing of steel had preceded Diesel's birth by two years. Germany's August Wilhelm von Hofmann's discovery of aniline in coal tar opened the way for aniline dyes and the mighty industries they represent. In 1861, when Diesel was three, the Belgian industrial chemist Ernest Solvay manufactured soda from salt; he was four when Sweden's Alfred Bernhard Nobel invented dynamite and began blasting his way to becoming one of the world's wealthiest men and, in time, one of the best-known philanthropists.

In 1873, as we have noted, Diesel's and Germany's own Professor Carl von Linde introduced ammonia refrigeration and with it complete independence from the rapidly vanishing natural ice. In 1886, the American Charles Martin Hall introduced the making of aluminum by electrolytic action, from which grew what is now the second-place metal industry of the world.

As a physicist, Diesel was appreciative of his predecessors and some of his contemporaries. His own father was still an apprentice leather worker when England's James Prescott Joule (in 1840) established a definite measurement for electric current; in Germany Hermann Ludwig Ferdinand von Helmholtz created the ophthalmoscope and in 1847 propounded the law of conservation of energy. Diesel was fifteen when the Scot James Clerk Maxwell suggested the electromagnetic theory of light. Only "last year" (1897) Scotland's Charles Thomson Rees Wilson had developed the "cloud chamber" for the detection of subatomic particles, while an Englishman, Sir Joseph John Thomson, was formally identifying the electrons.

In 1832, when Diesel's grandfather was chasing butterflies over verdant Bavarian hillsides, America's artistic Samuel Finley Breese Morse had perfected the electric telegraph, while the

German mathematician Karl Friedrich Gauss was outlining his system of measurements of electricity. Eight years later the American Charles Page invented the induction coil. Diesel was a precocious schoolboy of twelve when in 1870 the Belgian electrician Zénobe Théophile Gramme built the first practical dynamo. Diesel was a class leader in technical school in 1876 when the Scottish-American Alexander Graham Bell invented the telephone.

Progressively, Diesel was a Swiss factory worker when the English physicist William Crookes discovered the cathode ray, and an inventor-salesman of refrigeration machinery in 1885 when the American Henry Stanley invented the transformer for electrical currents. In 1892, when the Hungarian-born American electrician Nikola Tesla built the first alternating-current motor and Charles Proteus Steinmetz, the German-born American "electrical wizard," discovered the law of hysteresis as it affected alternating current, Diesel himself was even more renowned as a builder of engines. Two other notable contemporaries were the German Wilhelm Konrad Röntgen who discovered the X ray in 1895 and the Italian Guglielmo Marconi who in 1892 developed the first practical radio telegraphy apparatus.

From the America of the 1830's, Cyrus Hall McCormick's grain reaper and Samuel Colt's revolver came forth to change the basic economy and influence the morality of man. From Sweden, John Ericsson's screw propeller (1837) was an almost instant revolutionizer of international trade. In 1846, the American Richard March Hoe invented the rotary printing press and did much to make possible the mass production of newspapers, magazines, and books. The same year, the American Elias Howe perfected the sewing machine which had enabled Diesel's father to survive for a time as an upstairs manufacturer. In 1852, American Elisha Graves Otis improved the elevator by his safety brake and paved the way for building skyscrapers. In 1868, American Christopher Latham Shole's perfection of the typewriter (first conceived by Mitterhoffer in Austria) opened the door for women in "business." The following year American

George Westinghouse's completion of a practical air brake assured success for railroads.

Both as a student and as a neophyte inventor Diesel had watched with intense interest the beginnings of the motor-driven carriage. He was twenty-one when German Karl Benz built his first two-stroke-cycle engine, twenty-seven when Benz took his first ride in his "horseless carriage" powered by this engine with a differential gear. During the same year (1885) the German engineer Gottlieb Daimler installed his first gasoline engine on a motorcycle and the following year in his first "horseless carriage." Two years later Scotland's John Boyd Dunlop first put pneumatic tires on bicycles, thus making ready a juncture of the automobile and its primary accessory.

As a photography enthusiast, Diesel had viewed with special interest the work of French physicist Gabriel Lippmann in demonstrating his color-photography process in 1891. The myriad inventions of the American Thomas Alva Edison repeatedly attracted Diesel, beginning in 1877 with the talking machine, a year later the incandescent lamp, then the universal electric motor, the nickel-iron alkaline storage battery, and more recently (1893) the motion picture projector.

As time was to prove, Rudolf Diesel was well in advance of what may be described as the most inventive generation in the history of man. It was a generation, indeed quite substantially a century, of basic inventions on which the great industries of the twentieth century were founded. Many other historic firsts were in the offing, and several were vitally related to the strivings and anguish of Rudolf Diesel.

In another year (July, 1900) Diesel was to attend personally the successful launching of the first rigid dirigible of the Graf Ferdinand von Zeppelin at the Bodensee (Lake Constance). The 128-meter- (420-foot) long aircraft was powered by two Daimler engines. On first trial it flew "without incident" from Friedrichshafen to Immenstaad, a distance of five air miles. Count von Zeppelin suggested to Diesel that engines of such light weight as the Diesel engine were desirable to power his airship, but that they would have to be better proved operationally before he could use them. Thirty-five years later the giant airship *Hindenburg* was equipped with Diesel engines.

Meanwhile, still other uses were beginning to appear. Seven years earlier the American submarine inventor, John Philip Holland, had built and launched his first underwater boat. By 1899 the submarine was much more than a word. Naval architect Simon Lake was already expressing interest in powering his underwater craft with an oversize (600-horsepower) Diesel engine.

The future was bright. But the present remained dark.

A House, a Philosophy, and an International Success

RUDOLF DIESEL CONTINUED TO PONDER the plans for the first home he had ever owned. By 1897, he had bought a building site. It was a bold move, and even by the rising standards of the inventor's opulent thirties, it was an extravagant move. Munich of the 1890's was an old city with many sumptuous homes, the majority comparatively new. Some few of the handsomer ones had risen at the exorbitant cost of over 200,000 marks —by the prevalent exchange rates about $50,000 but in present-day purchasing power closer to $150,000.

Both Diesel and his wife gave exacting heed to the choice of a location. For a price they presently found what both agreed was a beautiful site on Maria-Theresia-Strasse, already an avenue of mansions—the word was *Villen.* The gradually lengthening line of élite homes faced a park of grass and trees, sloping down to the languid waters of the Isar River directly below. The place was suited for building a villa to excel all others. The asking price of the bare lot was fifty thousand marks. Diesel bought it.

Then began the two-year ordeal of house building. The inventor had told his friends, including the practical and ingenious Count von Zeppelin, that he intended to build a modest home on a suitable site.

The extent of building that followed did not sustain that statement. The excavation was comparable to that for one of the soon-to-be "flatiron" skyscrapers in America. The rising foundation showed a double wall designed to keep out dampness.

The overly deep cellar began to show characteristics of an indoor pavilion. It included a wide corridor in which sons Rudolf and Eugen rode their bicycles on days when the weather was bad.

Great stone piers marked the base sites for a number of oversize fireplaces. First sill work showed the outline of a vast two-story entrance hall presently to be paneled in oak and decorated by a fireplace inlaid with marble. Also planned were an immense living room and a Louis XV "salon." First outlines called for five marble-tiled bathrooms and an immense kitchen, which according to rumors, presently to be confirmed in fact, had been designed to be the most modern in all Munich.

Month after month the building went on. The huge, two-story entrance hall acquired a winding staircase with a banister carved from a single overlength oak log. All the principal rooms had fireplaces with hearths and mantles of marble. The five bathrooms materialized, all handsomely accoutered. The windows were "double openers" with novel frames permitting both sashes to be opened at once.

After a year of building, all the neighbors knew they were being outdone. Wonder after wonder continued to emerge. The dirt excavated from the immense cellar was being used as grade for formal gardens completely surrounding the new "palace."

As the furnishing of the house began, the symphony of opulence grew fortissimo. Almost all the rugs were Oriental or Persian. The entrance hall had a vaulted ceiling ornately carved and painted. Ponderous, hand-carved French Victorian furniture ballasted the great halls and the great rooms beyond and above. Mighty Renaissance-style wardrobes waited closely within. Busts, paintings, and miscellaneous art objects crowded the great hall. The garden room complete with high arched windows opened on formal gardens, which presently changed to a miniature arboretum-herbarium decorated with statuary and a high-spurting fountain. The dining room was provided with heavy mahogany furniture, bronze flower boxes, great palms, silk-upholstered sofas, and Persian rugs as wall hangings.

In the fantasia of real or imagined art works, Diesel quickly acquired a fondness for one particular painting. It was an evening scene on the Schelde River, painted by the contemporary Belgian artist Frans Courtens. In the foreground of the darkened waters, placid in the evening light, were two fishing boats. Frequently the inventor was seen standing in silence before the picture, gazing at it.

Martha Diesel took over the Louis XV salon as her very own. The walls were ivory, the rugs Persian, the writing table rococo; the total effect expensively, grandly feminine. The by now time-honored grand piano found quarters in the salon, but it was rarely played. Once dedicated pianist Diesel had somehow lost his devotion to the piano. On one occasion, so his younger son, Eugen, recalls, the inventor descended on the grand piano and played *"Allmächt'ge Jungfrau"* ("The Prayer of Elisabeth") from *Tannhäuser*. Although long out of practice, Diesel played brilliantly and with great feeling; at least it seemed so to his adolescent son. "We all gathered to listen to Father. His playing brought tears to all our eyes."

Diesel's personal room was an upstairs study furnished in the peculiar *Jugend* style—baroque ceiling, paneled walls with "skits" of handcarving, and a deep yellow Oriental rug. Over the brown marble fireplace hung a portrait of his wife, Martha. On the other side of the great room were picture windows facing the street, the gentle park, and the tired old river beyond.

The inventor had used many previous workrooms, ranging from corner cubbyholes to cluttered bedrooms and more recently a well-ordered apartment office. But this latest, the most expensive, was by far the least liked by its tenant. Yet the elegant study did have wall space for hanging engine drawings and self-addressed memoranda, as well as desk space for storing bills and discarded bankbooks.

The cost of building and furnishing the "modest" villa, which had turned out otherwise, was at least 700,000 marks. The total cost at the time of occupancy, including the cost of the site and other expenditures, was hardly less than 1,000,000 marks—about $250,000. Besides the cost of construction, there was the expense of maintaining the household. Servant wages alone represented a formidable burden of the overhead which was temporarily exceeding Diesel's income. It is extremely doubtful that Rudolf Diesel had one million marks at the time of the move (early 1900). His "side" investments had been bad and were rapidly growing worse. Bank account after bank account had been depleted and disregarded.

Martha Diesel's expenses added to the climbing overhead. Although the couple lived under the same roof, their relationship appeared to be growing more and more distant. They were no longer, to quote one of the inventor's phrases, "spending time together." Diesel worked in his personal room; his wife divided her time between supervising servants, keeping her rather complex social engagements (which were no longer her husband's), writing letters, and shopping in the city. In these pursuits and recreations she spent freely and traveled expensively, usually by hired horse phaeton or one of the swank brass-studded electric autos then to be rented (no extra charge for the driver) in Munich.

The children, as children will, were growing up. Daughter Hedy was being courted by a young automobile enthusiast whom she would presently marry. Young Rudolf, never robust and rarely even-tempered, was at art school in Frankfurt. When the father was away on business trips, Martha, her daughter, and younger son would assemble almost ritualistically in Eugen's austere room. Young Eugen felt keenly that his parents had drifted apart. This youngest child viewed with regret and a measure of distress their parting of interests. Martha was only a distant participant in her husband's multiplying problems. Rudolf appeared to be seeing more of Martha's portrait than of Martha.

Unquestionably Martha Flasche Diesel had long nurtured hopes of having a mansion. As a young wife she had established an admirable record as a helpmate by making the best of circumstances. In their modest Paris apartment she had lived quite happily and borne their three children. Residence in Berlin had created problems of adaptation, but as Berliners, Rudolf and Martha had been struggling up the social ladder together. Frau Rudolf Diesel had gloried, perhaps "basked" is a more accurate word, in the social prestige which her husband's work had created. Following their arrival in Munich, the inventor's colleagues, sponsors, and various other business and professional associates had been predominantly "important people," socially well established.

Martha Diesel knew very little about the intricacies of engines; and from what one gathers from her letters, she cared even less. But she did love her children. She also respected and still felt at least a degree of affection for her husband. During the weeks, months, and years of her busy and preoccupied mate's absences, Martha had been obliged to take over as head of the home.

The inventor's offspring were demanding youngsters, good looking, strong willed, never entirely easy to lead or, for that matter, to follow. Rudolf, junior, in particular, was a tense and aloof child, unusually sensitive and yet distinctly headstrong. The youngest, Eugen, though more pliable, was a "father worshiper." Almost from babyhood he had resolved to follow his father's lead, which was less than easy to do. In rearing her children, Martha regarded what she termed a place in society as their rightful heritage.

German wives of the late nineteenth century, like those of the early nineteenth, as a rule were neither expected nor freely permitted to participate in their husbands' business. Diesel's mother had actively helped her husband "keep factory"; but Elise Diesel had been and, for that matter, still was a decidedly exceptional German wife.

If Martha's interest in and acquaintance with her husband's career had dwindled and cooled, it was certainly not too surprising. Rudolf Diesel rarely confided in and almost never discussed financial affairs with his family. Only occasionally did he mention his technical problems or his toils with his customers or his strife with challengers and competitors. If Martha had all but ceased to offer her husband encouragement, she could and, in effect, she did answer that her husband had ceased to invite such encouragement.

As for the horrendous splurge on the palatial home which Diesel came to detest, Martha definitely wished to live in a mansion; she hungered for the attention gained by owning a luxurious residence. But Diesel himself had primarily conceived and brought about the building of what he shortly termed "this horrible mausoleum."

Perhaps in jest, perhaps more seriously, Diesel suggested that they sell the house, which was too large and too extravagant for the Diesels. Martha asked for time to consider the suggestion. Next morning she gave her verdict. She could not live without the house. "Is it necessary to sell it?" she asked. That ended the discussion.

Why had Diesel built the house? The inventor apparently never gave a decisive answer. A psychologist might juggle explanatory phrases, such as "resentment of childhood poverty," "desire to match or surpass the living status of professional friends and competitors," "artificial creation of stimulus and challenge," —that is, to live in and keep up the villa, the master simply could not fail or falter. Whatever the reason, the fact stands that building the Munich mansion overlooking the Isar was one of the more irrational acts of Diesel's life, and one of the most un-Diesel-like. As the inventor would presently admit and cry out, it was foremost among his most deeply regretted extravagances.

The mansion, now shorn of its one-time elegance, is a lingering monument to a paramount mistake—that it was built. Now the Diesel villa is just another government building peopled by minor bureaucrats, who, if not widely known or occupationally significant, are "world types."

The unhappy head of an unhappy expensive home, Rudolf Diesel responded to continued urges to be elsewhere. Haggard and weary from tense nerves and head pains now supplemented by foot pains (he had begun to suffer from gout), Diesel once more took the advice of his doctors to seek out a secluded place for a rest. Once more he journeyed to a temporary refuge, this time a rest home on Lake Constance in the Alpine foothills. His companion while there was his stimulating aristocratic friend, the Graf von Zeppelin, who lived in nearby Friedrichshafen.

The Lake Constance rest cure, which may not have been particularly restful and certainly brought no cure for the headaches or incipient gout, appears to have marked the debut, presently in print, of Rudolf Diesel as a social philosopher. It would be erroneous to imply that this development was either abrupt or represented an overnight change of intellectual direction. The inventor with typical thoroughness and refinement of details had been writing, collecting, and assembling notes for a treatise on the social responsibilities of the machine age.

The gist of the work was simply that for as long as he could recall, he had been deeply interested in ways and means of freeing working people from enslavement by the new machines, which obviously would multiply. The theme was implicit in Diesel's concept of a power unit suited to the work needs of the small shop as well as the large factory.

More care had been given to the machinery than to the people who had to attend it. As Diesel saw it, England in pioneering the Industrial Revolution had sought the fruits and profits of machinery without developing adequate machines to provide the power. The arms, legs, and backs of people, driven by the lash of poverty, if not by the rod or switch of the ever present overseer, had been forced to provide too much of the drive power. In Germany and other European countries where engine power had underwritten human muscle power more adequately, the machine was nevertheless enslaving its tenders with a heavy and galling yoke. Already factory machinery had accomplished the near slavery of what was being complacently termed the working class.

Diesel saw interdependence, "solidarity of interests," as the primary need for a machine age which could only continue and expand. He regarded the already bitter conflict of owner class *versus* working class as unnatural, unjust, and unendurable.

The only reasonable remedy, as the inventor saw it, was for

every worker to become part owner of the factory. The specific means which Diesel proposed was a penny-saving crusade wherein each shop or factory worker would deposit—in a glass jar or similarly convenient receptacle—a penny of his earnings for every day he worked. Thus, in a year, or two, or at most three, every worker could own at least one share in his employing firm.

To initiate the plan, Diesel placed on the handsome hallway table in his own mansion, and later on the dining room table, glass fruit jars for use as *"Spar-Büchsen."* Each member of his family and each servant or employee was invited to deposit a penny each day in readiness for the day when a share of the ownership would accrue to each one present.

If this share-the-ownership venture seems a bit ridiculous, it was at least a sincere expression of the inventor's concept of enlightened ownership. Diesel respected, ever so sincerely, the inherent dignity and "oneness" of man. As an observant traveler, he denied that one nation or race of man differs basically from another. He believed no less devotedly that mankind can live in peace.

Diesel was never a pacifist as such, though he sincerely believed that a society without war was attainable. Yet he did not object to existing militarism as "insurance against wars." Indeed, he continued to encourage the armies and navies of European nations to make additional use of his engines. He openly favored continuation of what he termed the "international capitalistic system," which would contribute to the international acceptance of his engine.

Diesel hoped that his ideas on the social aspects of industrialization would be forceful enough to start a reform movement. He called his book *"Solidarismus"* to emphasize the concept of social solidarity. The two-volume work did not find its way into print until 1903 although it apparently had begun taking form in 1899. The book turned out to be a notable worst seller; it sold a few hundred copies and was promptly forgotten by the general public.

At least in his own mind, Diesel succeeded in validating his devotion to peace, the removal of arbitrary boundaries, and man's liberation from the tyranny of machines and incarceration in social classes.

By gradual and painful stages the sensitive inventor found himself being brought face to face with the fact that substantial numbers of his fellow Germans and, indeed, people of other nations were not in a mood to seize on or long abide reform.

In particular, the politicians who led and fed from the fast-rising Social Democrat party spurned and, to a degree, belligerently opposed the Diesel suggestions. The politicians knew well that any successful move to abolish classes meant the abolition of the fastest-growing voting group and the most readily maneuverable. The rich man's friends desired to preserve and cajole wealth; the poor man's friends wished the laborer to remain dissatisfied with his lot.

For Diesel, these were bitter discoveries. At Maria-Theresia-Strasse 32, Munich, the inventor presently removed the penny jar from the great table. It hadn't collected many pennies. For that matter, neither for many months had Rudolf Diesel.

However, the inventor had solidified a dedication to work on with his engine, which Martha continued to refer to occasionally as his "black mistress."

Quite truthfully, the world's need for the Diesel engine was not imagined. The world entering the twentieth century needed engines in the worst way, and most experts agreed that Diesel's was the best engine. People required small engines for small shops; bigger engines for municipal and other public services and utilities; engines to pump water to the thirsting world; engines to help feed a hungry world; engines to power automobiles and small boats; huge engines to propel merchant ships and heavy naval craft; and, perhaps most important of all, locomotives to speed long trains on rails which would presently stripe every continent on earth. Rudolf Diesel had made and lost his first fortune. He proposed to recover by finding and filling more and bigger needs for better engines.

When Paris' International Exposition of 1900 finally closed

after having been "prolonged" by government decree until mid-November, the Diesel engine had won the *grand prix,* the highest honor bestowed on any exhibit. Interestingly, four of the five models exhibited had been built by the French-owned and engineered Bar-le-Duc corporation.

But 1900 was also a year of war, natural destruction, and want which resulted in demands for remedial measures and rebuilding. Great Britain, conquest bound, was floating a fifty-million-pound loan to finance the deplorable Boer War. The Boxer Rebellion exploded in China; nine Western powers were required to reseat the Bank of England in the Victorian superimposed ricksha. A ruinous hurricane ravaged its way across the Transvaal. Lower China suffered two of the most disastrous typhoons of the era. Cholera and famine simultaneously struck India; death tolls in Bombay alone reached fifteen thousand a week. Great earthquakes shook their way through Mexico, Venezuela, and Japan.

There was even more than usual need for public works for the good of suffering people. Rudolf Diesel, determined to do his part by means of his engine, met with a veritable maze of pathways ahead, beyond, and on all sides. For the next four years, the pain-plagued but extremely busy inventor followed the entanglement of trails. As he remarked to his sometimes merry friend, von Zeppelin, Diesel had "swallowed" the arrival of a new century in an infinitesimal fraction of one second, and was now finding himself obliged to spend four years digesting what had been so quickly gulped.

His efforts from 1900 to 1904 were largely directed toward Diesel engines and staving off bankruptcy. A survey made in May, 1902, showed a world total of 359 Diesel engines in use, developing 12,367 horsepower. The total number of problems was nearer 12,367 than 359.

In Budapest the new engine was the center of a storm of controversy. One of the principal consumers of electricity had bought and installed a Diesel engine to generate his own power supply. The substitution was as successful as it was economical.

He was thus able to produce his own current for one-fifth the rate the municipal supplier had been charging per kilowatt hour. But because it had lost its best customer, the municipally owned power company was obliged to raise rates. When the higher rates went into effect, the public charged the fantastic new engine with the horrid act of occasioning an unjust and destructive increase in power rates.

London, then the world's greatest metropolis, had no comparable complaint. The Diesel Engine Company of England, founded in 1900 with the inventor as a director, had built and sold one unit in the first year. By the fine old tradition of British brokerage, the British "department" had begun to purchase the engines from M.A.N. in Germany, Sulzers in Switzerland, Carels in Belgium, and others for resale all over the British Empire.

In France, a canal boat had been the first water craft to be powered by a Diesel engine. Early in 1903, the inventor personally and proudly rode the undersize barge, though by then it was no longer unique. Diesel's confidence in maritime use of what he continued to call his prime mover was soon to be confirmed by more impressive carriers. Already the Sulzer Brothers of Switzerland were outfitting both freight and passenger lake ships with two-cycle Diesel engines. By 1905, Carels Brothers of Belgium had built for exhibit (at the Liège world's fair) a 500-horsepower Diesel marine engine, at that time one of the biggest anywhere.

During 1902 the venerable little Amsterdam ship repair firm with a ponderously big name, *Nederlandsche Fabriek van Werktuigen en Spoorweg-Materiel,* acquired a license to build the new engine. Although founded to repair steam engines for ships, the firm had long before shifted over to designing and building new marine engines. Inaugurating Diesel power involved patent waits and other delays, but within six years the Dutch firm had made an impressive mark in ocean-craft history by building two enclosed four-cylinder 160-horsepower engines and installing them on two former steamer clippers, the *San Antonio* and *Cornelius.*

During 1903, Franco Tosi of Legnano in Italy acquired a license to build the engines. Within four years the first Italian-built unit was ready for sale.

Sweden was among the first with the new engines. During January, 1898, M. Wallenberg and O. Lamm acquired exclusive manufacturing and sales rights for Diesel engines and the following June transferred them to the newly founded *Aktiebolag Diesels-Motorer* of Stockholm. The construction of Diesel engines in Sweden according to Augsburg specifications began in January, 1899; by July of that year the first Swedish-made Diesel engine had been delivered to Swedish State Railways. It was a 20-horsepower model used for a pumping plant in Liljeholmen. Within another six months the Swedish firm had developed its first double-cylinder engine. By August, 1901, the *Diesels-Motorer* had succeeded in reducing by half the still excessive weight of one of the trunk-piston engines. Its fuel consumption of 193 grams per horsepower hour remained a record low for several years. With this innovation, the Swedish company made an important contribution to the advancement of Diesel engine design.

Many significant improvements were made later by Sweden's K. J. Hesselman, who also designed a simplified direct-reversible marine engine. A two-stroke-cycle Diesel engine was first installed in a sea-going vessel in 1907, and the first motor ship powered by Diesel engines crossed the Atlantic in 1911. The company, now known as the *Aktiebolaget* Atlas Diesel, also built the 180-horsepower Diesel engine which powered the polar-prying ship *Fram,* which carried Captain Roald Amundsen on his Antarctic expedition in 1911. After the Norwegian explorer succeeded in arriving at a point near the South Pole with his four companions and fifty-two dogs, the type of Diesel engine which powered the exceptionally good ship *Fram* was named the "Polar."

Additional Scandinavian progress in building the new engines was being reported from Copenhagen from the shipbuilding firm of *Aktieselskabet* Burmeister and Wain. In 1854 the veteran shipbuilders, then Baumgarten and Burmeister, had delivered a wooden steamer, the *Hermod*, to the Danish postal authorities. Soon after the American Civil War, the same company, by then Burmeister and Wain, gained widespread prominence as bridge builders with the famous union bridge connecting Copenhagen and Christianshaven. Meanwhile the firm's shipyards, grown to be the busiest in Denmark, specialized in de luxe seagoing craft. The shipbuilders had launched the Royal Danish yacht *Danneborg* and, in 1895, the palatial Imperial Russian yacht *Standard*, 5,255 tons, reputedly the handsomest ship afloat.

Late in 1897, Diesel had personally signed a licensing agreement for the manufacture and sale of his engine by Burmeister and Wain. The firm promptly built a model engine according to furnished plans but elected to stand by for tests and improvements of subsequent output. By 1903, the Danish firm began its own manufacture of Diesel engines, based on the design developed by the Swedish license holders but substituting certain features of the original Augsburg designs. During 1904, Burmeister and Wain completed the building of its first ten Diesel engines. These ranged in horsepower from 8 to 160 and included both the one- and two-cylinder units. Most of these engines were installed as power plants. After sixty years several remain in service, proof extraordinary of excellent workmanship and superior design.

Quite correctly the inventor sensed that the Danish Diesels would prove to be of historic consequence. In 1909, the company entered the marine engine field with their Diesel engines. Their first commission, from the East Asiatic Company, was to construct the *Selandia*, the first major-tonnage cargo vessel to be powered entirely by Diesel engines. Long before the first vessel was completed, East Asiatic had placed orders for two others, the *Fionia* and *Jutlandia*.

The two four-stroke-cycle, eight-cylinder, direct-reversible Diesel engines which comprised the *Selandia*'s power plant had a piston diameter of 530 millimeters (20.87 inches) and at 140 revolutions per minute developed 1,250 horsepower each. Relieving were two auxiliary Diesel engines of 250 horsepower each (at 230 rpm). With this force, reversing from full speed ahead to full speed astern could be carried out in less than twenty seconds. The average speed of the substantial 4,950-ton vessel was eleven knots, as compared with four to seven knots for comparable steam craft.

The *Selandia* set out on her maiden voyage to the Far East on February 22, 1912. To give the English people an opportunity to see "this new marvel," the ship docked at the West India docks at London. During her visit there many distinguished visitors came aboard.

The First Lord of the Admiralty Winston Churchill, who later proved that he was not invariably wrong, said, "This new type vessel is the most perfect maritime masterpiece of the century." Later, when His Majesty's First Lord was asked in the House of Commons what the policy of the Admiralty was regarding this new, advantageous method of propulsion, Churchill replied that he had already given instructions to investigate the possibility of employing Diesel engines for use in British warships.

The *Selandia* arrived in Bangkok on April 20, 1912. The engines were overhauled after running 980 hours. The ship weighed anchor on May 1 and arrived on June 26 in Copenhagen, amidst "tumultuous jubilation." The 22,000-mile maiden voyage had been successful in every respect.

After the engine had been in service more than twelve years and 600,000 miles, the chief engineer of the *Selandia* wrote in his report: "During the past twelve years we have spent ten days in port. . . . We have often returned from Bangkok and after one or two days' stay in Copenhagen started on the long voyage out to the East again."

The sister ship, *Fionia,* made her maiden voyage to Kiel, where Lord Pirrie, managing director of Harland and Wolff, and Herr Ballin, director of the Hamburg American line, were attending the regattas. The shiny new ship without funnels attracted attention as it glided, invisibly powered, past the gaily decorated ships of the German fleet into the harbor. The *Fionia* made such an impression on Ballin that he insisted on buying her for his line instantly. Terms were arranged, and the house flag of the East India Company exchanged for that of the Hamburg American line. The original owners insisted on rechristening the ship *Christian X* in honor of the reigning king of Denmark. While these arrangements were being made, the German Kaiser, Wilhelm II and a large party inspected the vessel.

After many other successful marine installations, Burmeister and Wain built a twin-screw Diesel-powered plant for the 19,000-ton motorship *Gripsholm* of the Swedish American line. During the next few years the firm and its associates supplied more than twenty similar Diesel engines for large passenger liners. Little Denmark was indeed foremost in taking Diesel engines over the seven seas.

The Black Mistress Goes International

⚏
⚏

⚏ For better or worse, Rudolf Diesel was strongly established as an authority on paradoxes.

He had confronted or created most of the principal kinds of paradoxes—personal, professional, legalistic, domestic, parental, and ancestral. Now, as the inventor grew ever more determined to devote what remained of his talents and his life to seeing the engine developed into a global asset, he had reason for pondering the paradox of acceptance of his engine in principle but delay in making practical use of it in commerce.

On his better days—those of undisturbed spirits and less than usual pain from his headaches and his gout, which by now was beginning to make necessary a specially built shoe or slipper for his right foot—Diesel strove to rationalize calmly. The paradox of theoretical acceptance but prolonged deferment of wide distribution of the Diesel engine would have been tantalizing even for a robust inventor with Jobian patience. That the engine had a great future required no further proof. It was already a necessity of industry. Diesel maintained his steadfast resolve to keep on improving and adapting the rational heat engine rather than to fritter away his time, energy, and talent on lesser inventions, even though they might have turned out to be profitable.

The resolution was not easy to keep. There was, for example, the expanding field of mechanical refrigeration in which Diesel had already achieved eminence. As a thermal engineer of accredited stature, he could have returned to work with what he termed the low end of the mercury column promptly and profitably.

But Diesel had a deep-seated, partly mystical devotion to the engine which his wife more and more frequently referred to as his black mistress. He also was aware of the many voices that were crying out from the new industrial wilderness. There was need, ready-made, crucial, and unmet, for more adaptability and greater efficiency in engines. Diesel knew he had the basic answers for this need. At forty he saw his engine both as fulfillment of a global need and as a very special Rubicon, from which he could not or would not retreat. Even forgetting the requirements of his homeland, which the inventor had no desire to do, he felt responsible for the development of the machine which he had created so that it could serve the whole world.

Imperial Russia had made a memorable beginning with the Diesel engine, through Emanuel Nobel. Ready and waiting for it were the Ludwig Nobel machine factories of St. Petersburg, supported by the Nobels' Galician oil fields, already richly productive and thus in the market for petroleum-fueled power-developing machines. Also waiting were the Nobels' fleet of oil tankers and other commercial shipping, all in need of more efficient and faster propulsion.

Emanuel Nobel had already proved good faith. He had paid 600,000 marks in cash plus 200,000 marks in stock certificates, the latter directly to Diesel, for the exclusive right to introduce and supply the Diesel engine to all Russia. The immigrant from Stockholm intended to demonstrate that he had spent wisely.

The first Diesel engine built in Russia, in 1899, was a modest-sized pilot model. The following year the Nobel factory in St. Petersburg built three Diesel engines of about one hundred horsepower each for the oil-pumping operations in Baku, the long-lived oil center of what is now the Azerbaijan Republic on the west coast of the Caspian Sea.

In October, 1902, Nobel licensed the machine factory at Kolomna near present Moscow to build Diesel engines to help supply the great demand of the oil industry. During the follow-

ing year, fifty-two 150-horsepower Diesel engines were manufactured for powering the pipelines from Baku to Batum, capital
of the now Georgian Republic of Adzhar, at the foot of the
Caucasus Mountains on the Black Sea.

During 1903, Nobel's merchant steamer *Wandal* was equipped
with three three-cylinder, four-stroke-cycle marine Diesels built
by *Diesels-Motorer* of Stockholm. Since the Diesel engines could
not be reversed, each of the twin screws of the vessel was driven
by a 75-kilowatt electric motor. Fuel cost for the *Wandal,* which
plied both the Volga River and the Caspian Sea, was only 20
per cent of that for steam-powered vessels. It provided one of
the first and most impressive demonstrations of the merits of
the new engine in marine use.

During 1904, Nobel's Russian-chartered oil tanker *Ssarmat*
was also equipped with Diesel engines, built by the licensee in
St. Petersburg. By then, and to the inventor's grateful surprise,
the Diesel engine was off to its fastest start in Russia.

France, too, was contributing to the progress of the "wonder
engine." Duly licensed by the *Société Française des Moteurs
Diesel,* the *Anciens Établissements* Sautter-Harlé of Paris began
manufacturing during 1899. Sautter-Harlé specialized in marine
engines but extended its production to include Diesel engines
for powering electric generators. Beginning in 1903, the Paris
firm worked to develop a reversible Diesel engine. That led to
building a "lie down" (horizontal) opposed-piston marine engine of 120-horsepower at 400 revolutions per minute. Two such
engines were installed in the famed pioneer French submarine Z.
A 25-horsepower Diesel engine of similar construction was installed in the canal boat *Petit Pierre* about the same time.

In 1904, Sautter-Harlé built four-cylinder, four-stroke-cycle
Diesel engines of 300-horsepower (at 340 rpm) for three other
submarines of the French Navy. In the same year it produced a
four-cylinder experimental engine with solid fuel injection
capable of higher speed. This engine was installed in a motor-
boat. But the most important French contribution was adapting
the Diesel engine for submarine use. By 1907, Sautter-Harlé was

building much larger Diesel engines (of six cylinders, generating 700-horsepower at 400 rpm). The 1907 model was the horizontal type with air-cooled pistons and a new kind of scavenging system designed to utilize the exhaust gases ordinarily wasted. Later Sautter-Harlé built the first all-enclosed four-cylinder submarine Diesel engines.

Somewhat to the inventor's surprise—certainly he expected no such outcome during his first rather grim visit to Scotland— the Scottish clansmen were also producing engines. Late in 1898 the veteran Glasgow firm of Mirrlees, Watson and Yaryan Company had turned out the first operative Diesel engine built in Great Britain. It was a single-cylinder, 50-horsepower unit which is now on exhibit in London's South Kensington Science Museum. It was placed there after nearly half-a-century of useful work.

After some faltering and discreet gasping about the formidable costs of manufacturing the newfangled engines, the Scottish company (reorganized in 1900 as Mirrlees, Watson, Ltd.) acquired a new manager and began an impressive spurt of progress in making the Diesel engine practicable for fulfilling needs for less weight and more speed. Mirrlees also succeeded in building larger units powered at about fifty horsepower per cylinder. By 1903, the British Admiralty placed an order for two three-cylinder Diesels and one four-cylinder, all for installation as auxiliary power units for the major battleship H.M.S. *Dreadnaught,* the first British naval craft to be equipped with the "wonder engines."

Three years later the Mirrlees engineers designed, on Admiralty orders, lighter-weight engines for powering subsidiary craft, such as pinnaces, or "long boats." These were fast-running, four-cylinder engines, developing 120 horsepower at 400 rpm. They had aluminum bedplates and columns of manganese bronze. After fifty years they are still impressively "modern." To the surprise of many, one suspects Diesel included, Scottish engine-building was moving forward rapidly.

As the inventor had definitely expected, the Sulzer Brothers of Switzerland was also making significant progress. Since his initiation into factory work at Winterthur in 1879 when he was only twenty-one, Diesel had thought of the Swiss concern as his alma mater. First opened as a brass factory just as the American Revolution began (the founder bore the alliterative name of Solomon Sulzer), the Swiss firm had continued as an especially aggressive and inventive concern.

As already noted, in May, 1893, Diesel had made a provisional agreement with Sulzers which granted the firm Swiss patent rights for the new engine and provided that it should build an experimental engine. After a discreet wait to see the opening move of the Augsburg-Krupp combine, the Swiss brethren completed their first Diesel engine in May of 1898. It consisted of a single upright cylinder with a 10.24-inch (260 millimeters) bore, a 16.14-inch (410 millimeters) stroke, and a power development of 20 horsepower at 160 rpm.

Sulzers did not place this first model for immediate sale. For many years the firm had been manufacturing large steam engines mostly for ships. It saw its principal advantage in larger-unit Diesels for marine use.

On receiving an order from the Diesel Engine Company of London for twelve Diesel engines of thirty-five horsepower each for resale in England, Sulzer Brothers began serious manufacture of the machines. The company then confirmed its exclusive ownership of the Diesel patents in Switzerland with the right to export engines to all countries of the world for a period of twelve years. The same agreement provided for a free exchange of technical improvements with Diesel and all his license holders.

Under the guidance of Jakob Sulzer-Imhoof, director of the newly established Diesel department at Sulzers, a standard Diesel engine developing forty horsepower per cylinder was soon built in sizes of one-, two-, and three-cylinder units. Within three years twelve different cylinder sizes were manufactured with outputs of from twenty to two hundred horsepower. The first models were four-stroke-cycle engines of the trunk piston type without piston cooling. The fuel was blown into the cylinders

and atomized by compressed air. From these ingenious beginnings the Swiss firm continued making Diesel engines of
excellence and typical durability.

One engine delivered in 1904 to Port Said, Egypt, and another
to the Benedictine Monastery at Einsiedeln, Switzerland, were
still in effective operation more than forty years later. Sulzers
also moved ably into the manufacturing of multi-cylinder engines with greatly increased power development. Within another
decade, the firm was producing four-cylinder Diesel engines with
capabilities of one hundred horsepower, and shortly thereafter
other assemblies with outputs of six hundred horsepower per
cylinder. These developments foretold the power and importance to come.

Meanwhile, Rudolf Diesel sensed that the United States might
well become the main customer for his chief invention. His
confidence in what he had termed "the great light to the west"
was growing steadily. In considerable part that confidence was
associated with the leading beer man of St. Louis.

Adolphus Busch, in turn, was well satisfied with his acquisition
of exclusive American and Canadian rights to manufacture and
sell the Diesel engine. Because he was a shrewd businessman,
the beer king had not yet fulfilled all terms of his important and
costly contract with the persuasive Diesel. Busch had grown fond
of the inventor, and his friendship was cordially, if somewhat
shyly, returned. Herr Busch owned the fashionable villa "Lilly,"
in Lagenschwalbach near Frankfurt and a nearby hunting lodge.
He spent several months of every year at the villa and the lodge,
in Munich, and at watering and health resorts, where he continued from time to time to meet Rudolf Diesel.

The guardian angel of the long-distance friendship was the
diligent technical adviser to Adolphus Busch, Colonel Edward
Daniel Meier.

Although born and reared in St. Louis, Meier was German
educated. For four fruitful years he had studied engineering at
the Polytechnic Institute at Hannover. After that the serious-
miened young adventurer had returned to the United States,

enlisted in the army, and served for three years with the Far West Indian Patrol. With the outbreak of the Civil War, he transferred to McClellan's First Army and "fought his way up" to the rank of colonel. After being mustered out, Meier found employment as a mechanical engineer in the locomotive works at Paterson, New Jersey. Soon he became superintendent of the Kansas Pacific Railroad, forerunner of the Union Pacific and employer of showman and hunter William F. (Buffalo Bill) Cody. In 1884, Meier founded the Heine Safety Boiler Company in St. Louis and presently became prominent in the manufacture of boilers and various heavy machinery.

Later, "Colonel Ed" was made president of the American Society of Mechanical Engineers. More pertinent in terms of Diesel engines, the Colonel was in close touch with the manager of engine construction at the *Maschinenbau-Aktiengesellschaft* Nürnberg, the company which had undertaken the construction of a Diesel engine under license from the *Maschinenfabrik* Augsburg.

Some five years before his first meeting with Busch, Diesel had submitted to the United States Patent Office a lengthy description of his "method and apparatus for converting heat into work" with detailed drawings. On July 16, 1895, after a wait of nearly three years, the inventor received United States Patent No. 542,846.

Busch had talked with the Krupp directors and engineers about the Diesel engine and had visited the Augsburg factory before sending Meier to Augsburg to inspect the engine. Meier responded with a thorough job of investigation and appraisal. The closing paragraph of his detailed report is a revealing look at industrial estimates of the times:

"I believe that the purchase of the Diesel patents for America is as promising an investment as the purchase of any patent claim could be. Applying the same methods of giving licenses in the United States, which Mr. Diesel has so successfully used in Europe, we should soon create from among the largest and best engine works of the United States an interest so strong that no one would dare to attempt infringement or evasion of our patent. Every practical builder will think too highly of the value of the long experience of the Augsburg Works in developing the first working Diesel motor, to think of trying to do without them at the risk of a patent suit."

The contract between Adolphus Busch and Rudolf Diesel, written in German and containing eighteen separate clauses, was signed by both parties on October 9, 1897, in Munich. The Canadian patent, No. 44,611 of November 3, 1893, was also involved in the transaction; it was penned in longhand into the typewritten contract, apparently as an afterthought. At the moment of signing, Busch handed Diesel his check for 1,000,000 marks in gold, $238,000 at the current rate of exchange.

On New Year's Day, 1898, the Diesel Motor Company of America was formally launched with offices at 11 Broadway in New York. Adolphus Busch was president; his son-in-law, Hugo Reisinger, was business manager; Edward D. Meier, technical director; and Arthur F. Frith, chief designer, with Hugh Meier (nephew of Edward Meier) and Edward Puchta, assistants.

Since the new company did not have facilities to manufacture engines, it was decided to license selected machine works which were so equipped. Two single-cylinder, 20-horsepower Diesel engines had been ordered from the Nürnberg and Deutz factories to serve as demonstration models. Hugh Meier and Puchta went to Germany to learn at firsthand the various construction processes and to accept delivery of the model engines. Meanwhile, Colonel Meier set about giving lectures on the Diesel engine before technical groups and writing explanatory articles for magazines and newspapers.

Late in March, 1898, the engine built by the Nürnberg factory arrived in New York. The German engineer, Anton Böttcher, followed in April to assist in installing the machine and in making the first tests. The engine built by the Deutz works did not arrive in New York until July. Meanwhile, the first engine had been installed in a laboratory near the company's office and rigged with a 15-kilowatt direct-current dynamo manufactured by the C. and C. Electric Company of Garwood, New York.

On April 23, 1898, a Diesel engine ran for the first time in America. It promptly blew a piston. Herr Böttcher had rather expected that and had brought along appropriate spare parts. Duly repaired, the wonder engine was set up for the opening of the electrical exhibition at old Madison Square Garden on May 7.

The new company, meanwhile, had prepared its first advertising promotion, a four-page circular carrying a picture of the engine on the front page with the caption: "The Diesel Motor. The Rational Heat Engine. The Source of Power for the Future. Exhibited for the First Time in the United States at The Electrical Exposition at Madison Square Garden, May 1898."

The selling text was more emphatic than any the inventor had offered: "This New Heat Engine is the invention of Mr. Rudolf Diesel, an eminent and scientific engineer of Munich, Germany. After 15 years of exhaustive study and careful progressive experimenting, he has given to the engineering world a practical machine with Double the Economy of the most perfect triple expansion steam engine and 50 percent in actual efficiency Above the hitherto best gas or oil engine.

"The best results of theory and design in steam and gas engine practice have been utilized in its construction, and the Dream of Sadi Carnot has been shown capable of realization to an extent heretofore considered impossible . . . German theory and study have given this demonstration. American skill will adapt it to all uses and place it within the reach of all who use power!"

"This America," Rudolf Diesel reflected, "may very well be the haven and heaven for all inventors."

Diesel's interests might range far, but they always came back to *Maschinenfabrik* Augsburg, A.G. Here the first marketable Diesel engine had come into being—a two-cylinder, 60-horsepower model sold to the match factory union at Kempten. And, in December, 1893, the first engine-making merger, M.A.N. (*Maschinenfabrik* Augsburg-Nürnberg) had come about.

M.A.N. scored its first notable commercial success in 1902 when its now classic A-frame engine was purchased by the city government of Kiev, Russia, for generating electricity to operate a streetcar system. The streetcar administration of Kiev ordered four Diesel engines with a total output of sixteen hundred horsepower. When the contractor who supplied the electrical equipment for the plant refused to guarantee the satisfactory operation of the Diesel engines, the M.A.N. factory took over the guarantee for the entire installation. Subsequently, two additional engines of identical construction were added. This gave the Russian city what remained for many years the world's largest operating unit of Diesel power.

A primary cause of Russian acceptance of the new engine was her abundance of inexpensive fuel oil. In Germany, on the other hand, the fuel situation was highly unfavorable, since the German petroleum dealers were still paying a 40 per cent license fee to the American "oil trust," their sole supplier. The Russian Alexander Iswolski pressed for the speedy completion of the Danube waterway as a solution to transportation difficulties, so that oil from the rich Russian fields could be brought into the German market to compete with fuel from America.

Diesel was well aware that the success of his engine from the standpoint of numbers in use depended basically on fuel supply and steel supply. The latter was a crucial selling point in his original negotiation with the Fried. Krupp *Werke* of Essen. Diesel had submitted his patent drawings and descriptions to Friedrich Alfred Krupp with the open suggestion that since his proposed engine was to be constructed principally of steel, the foremost steel firm, which in its Krupp-Gruson plant was already manufacturing gas engines, could profitably correlate quality steel production and manufacture of Diesel engines.

Friedrich Krupp had promptly communicated with the Augsburg *Maschinenfabrik* regarding its claim to the Diesel patents and reviewed the prospects with his own *Grusonwerk* at Magdeburg-Buckau. The resulting contract, dated February 25, 1893, gave the Krupp works sales rights for all German territories not already ceded to the Augsburg concern, plus those of all Austria-Hungary, where Arthur Krupp's Berndofer *Metallwarenfabrik* was firmly established. The initial contract terms had been highly favorable for the inventor: a payment of thirty thousand marks yearly during the initial testing period and a royalty payment of 37.5 per cent of factory prices of engines actually produced and sold until Diesel had received 500,000 marks. Thereafter, he would receive royalty payments of 25 per cent of the net selling price. Costs of developing a marketable Diesel engine were to be borne equally by the Krupp works and the Augsburg machine factory; the resulting profits were to be divided equally by the three parties: Krupp, Augsburg, and Diesel. The Krupp contract included rights for stationary Diesel engines and for automotive, marine, and locomotive units as well. Following formal signing of the contract, the Augsburg-Krupp combine was founded on April 25, 1893, for the construction of "a practical Diesel engine."

Diesel had suggested that Krupp undertake tests with fuel gases of various types as then available to the Essen factory, generator gas, illuminating gas, and water gas. The Augsburg factory, meanwhile, undertook the testing with liquid fuels.

The Krupp engineering staff had prodded Diesel to speed up the complex and inevitably expensive testing processes and had repeatedly shown annoyance with what he deemed necessary or unavoidable delays in the laboratory work at Augsburg. Yet after four years, when the *Krupp-Grusonwerk* began the commercial manufacture of a 20-horsepower Diesel engine (in June, 1897), the great steel maker completed only one engine in the first six months and then devoted the next seven years to "further study."

But Krupp had remained active, at least with regard to experimentation. During 1898, the firm had exhibited its so-called Hartwig model at the Munich industrial fair, and, thereafter, put the engine in service in the Germania shipyard at Kiel. The Krupp development of an improved air-pressure tank for use in starting the engine and injecting fuel was of substantial value. Krupp fell behind, however, as an effective builder of engines for commercial sale. The biggest iron-steel-coal manufacturing complex preferred engines which could be fueled by coal products. Although the Gruson plant had developed a 300-horsepower, one-cylinder combustion engine to operate on coal gas, there still seemed to be no practical assurance of a coal-burning Diesel engine. The Fried. Krupp *Werke* simply was not interested, at that time, in manufacturing an oil-burning Diesel engine.

To aggravate the delays, in June, 1898, fire destroyed the gas-engine manufacturing facilities of Krupp's Gruson plant. Casualties included several Diesel engines under experimental construction. Since the department had operated at a loss for many years, the management decided not to rebuild. However, two partly built 20-horsepower Diesel engines were saved and forwarded to the Essen factory for completion. They were installed in the locomotive shops.

A ray of hope appeared when Krupp directors requested their Berlin-Tegel works to study the adaptability of the Diesel engine for marine use. In 1896, Krupp had acquired the *Schiffs- und Maschinenbau-Aktiengesellschaft Germania,* which included ship-building yards at Kiel-Gaarden and a machine factory at Berlin-Tegel. Shop drawings of a 20-horsepower engine were sent to the Berlin works in 1898. Early the following year, on orders of *Maschinenbaudirektor* Schultz of the Krupp Berlin works, construction of a four-stroke-cycle, four-cylinder marine Diesel engine of sixty-five to seventy-five horsepower was scheduled to begin.

But the plans clashed with the syndicate's (*Allgemeine Gesellschaft's*) plans for a two-stroke-cycle marine engine which Chief Engineer Güldner insisted would better serve the requirements of ship propulsion.

The proposed two-stroke-cycle, three-cylinder marine Diesel engine (fifty horsepower at three hundred revolutions per minute) was never built. At least for the time Krupp was skittish about expanding its investments in Diesel power at sea or, for that matter, on land.

The decision was only a delaying tactic, for Krupp eventually took its place as one of the important manufacturers of the engine. But the interim verdict exemplified a dilemma which Diesel faced as the old century ended and the new one began. He already had received his larger and more luscious financial plums. Krupp's agreement to pay royalties on the sale of the engines added up to zero, which was no comfort at all to the sizable Diesel living expenses. Furthermore, big business was disposed not to take on the "radical" innovation, while, in many instances, small business lacked the capital, credit, and technical skills to do so. Even so, Germany provided a majority of both first sponsors and able competitors.

In 1892, when his engine was newly patented, Diesel had tried aggressively to interest the greatest potential competitor, Deutz of Cologne (*Gasmotorenfabrik*), in manufacturing his engine. Deutz then led the field in the manufacture and sale of small engines fueled with gas, and Diesel respected the experience and testing techniques of the firm. One of the Deutz company's founders was an unusual salesman, Nikolaus August Otto, who had invented the gas engine.

Although lacking technical schooling or experience, Otto, in 1861, had devised and built the first practical four-stroke-cycle internal-combustion engine. The following year he built a successful four-cylinder gas engine. Financier Eugen Langen then joined Otto in founding *N. A. Otto & Cie. Fabrik Atmosphärischer Gaskraftmaschinen*. During 1866 they succeeded in patenting their engine in Germany, England, France, and other countries. At the Paris world's fair in 1867, the Otto gas engine proved to be a sensation; it won the gold medal, the highest award of the exposition. Thus encouraged, the firm in 1872 built a factory in Deutz, *Gasmotorenfabrik* Deutz, with a capital of 300,000 taler (some $214,000) and with Gottlieb Daimler, soon to emerge as an automobile pioneer, as works director.

During 1875 the company developed its first gasoline engine. In May of the following year, its first four-stroke-cycle gasoline engine ran successfully and was put into production. The next important step was patenting and manufacturing a magneto ignition system. The company, already busy producing salable engines, at first rejected the Diesel overtures. The Deutz men had learned the hard way the tremendous amount of work and capital required for developing any new type of engine.

In 1897, however, having watched the lengthy experiments of Diesel and the Augsburg-Krupp combine, the Deutz works purchased a license to build Diesel engines. Their first engine was shipped to the American Diesel Motor Company in 1898. The second machine was exhibited at the Second Power and Works Machine Exhibition at Munich in the same year and was later used in their own factory.

Even so, the Deutz directors still believed that the Diesel engine was too complicated for the prevailing market. It did not fit their basic objective, the production of a simplified engine; it just didn't fit the Deutz slogan, "Simplified Engines." On second thought, Deutz decided to simplify the Diesel engine with its own toil and money. The firm designed three models of Diesel's engine, but completed only four marketable units; then for a period of six years suspended their manufacture.

The eventual revival was quite historic, because it saw the development, first, of an upright engine with direct air injection; then, of a horizontal engine, followed by a new type of injection system without benefit or handicap of air compressor; and finally, an engine with a pre-combustion chamber. The special hero of these feats was a laboratory engineer with the engaging name Prosper L'Orange.

Diesel could and did admit that his competitors-turned-colleagues were proving particularly effective. But the engine was gaining manufacturers on all sides. As 1901 ended, a total of thirty-one responsible companies were licensed to build and sell Diesel engines. On paper, sales franchises covered eleven nations, with many more to follow.

Yet Diesel engines were not selling in great numbers. As 1902 ended, fewer than 400 were in use. During the following year the total had climbed to 463 known and accounted for. The inventor reiterated his conviction that the returns would presently come sweeping in like "an argosy from afar." That statement was more than just another poetic Diesel simile. The inventor was confident that in the almost visible future his engine would power most of the great naval and merchant fleets of the world. No other prime mover, no other rational heat engine could propel seacraft nearly as efficiently or dependably.

He was no less confident that medium-sized Diesel engines would power railroad trains on every continent. Electricity would presently be hardly more than a by-product of the same genus, the Engine Diesel. With near clairvoyance the inventor previewed the odyssey of the age of automobile. He was quite certain, too, of the kind of engines which would presently propel all, or practically all, automobiles and trucks. They, too, would be Diesel engines—made smaller.

When the Austrians Professor Ludwig Czischek and wagon manufacturer Ludwig Lohner came to Munich on August 24, 1897, to seek Diesel's advice about an engine for their entry in the 1898 "reliability run," the inventor could not promise an engine suited for automotive use the coming spring, but he said, "I am convinced that the particular peculiarities of my engine will achieve victory over all others, especially in the automobile industry."

As he hurried into middle age, Diesel remained wholly convinced that his own destiny remained with that of his engine, further developed and perfected for the good of and service to all mankind and its world. "If the whole thing has a purpose, if people are made happier by it . . . ." The inventor ended the monologue to his youngest son with anti-climax: "But that I would not care to decide any more today . . . ."

With minimum delay, he began converting the entire upper floor of his mansion in Munich into a central construction office.

In room after room he replaced the expensive furniture with long drawing tables and draftsman's tools. The move seemed grimly defensive. "I can't endure this outrageous mausoleum," he cried out. "It is making me a dismal blue corpse!" Once more he would work his way to resurrection.

Instead of tea guests, recruited so tirelessly by his wife, Diesel began filling the top floor with draftsmen and consulting engineers. His days at home were spent in shaping a combined design center and planning laboratory. The high, luxuriously paneled walls were blotched with blueprints and scale drawings of unborn engines and their parts. "My upper story," Martha Diesel told her sister-in-law, Emma, "has come to look and smell like the drying room of a tannery."

The wall displays were destined to change history. Some were hardly more than fantasies. Many were sketches of engine designs which even now would be regarded as far ahead of the times. There were projected models of marine engines for propelling all sorts of watercraft—from tiny fishing boats to transoceanic liners many times the size of any yet built. There were designs applicable to aircraft and underwater craft and automobiles, all of which the inventor found enormously challenging as ideas already incubating. His journal notes of this period reveal Diesel's tremendous perceptiveness: the nineteenth century had been the century of earth-changing, life-changing, basic inventions; the twentieth century would be for developing the great formative inventions made during the previous hundred years. The thesis remains. Most patents nowadays are for progressive and functional improvements—not for basic or primary inventions. The groundwork had been solidly laid.

In his magnificent home laboratory, Diesel, with the able help of Heinrich Deschamp, planned a small engine suited to powering horseless carriages. Again the immediate reward was in recognition rather than in money. The "petite model," as he called it, received a *grand prix* at the Brussels world's fair in 1910, though the expert consensus was that the award was primarily in recognition of prior work in developing engines rather than specifically for the little one. Regardless of the award, very few of the first five-horsepower engines were sold.

In the newest and most de luxe of his home workrooms, Diesel not only planned but also assembled a four-cylinder automobile engine and saw it installed in a van chassis, appropriately named "Van Vou." But once more that ugly word "practicality" intervened. The proposed automobile engine had several operational faults. It was far too bulky even for present-day automotive use and was cluttered with the most inconvenient gadgets for starting and for air compression.

Aside from the technical faults, there was the expense of constructing the engine. The automotive field was already sharply competitive, and because of the necessarily high grade of metals used and the precise tooling and shop work required, the Diesel engine remained far more expensive than gasoline engines. The ratio of three to one was approximately standard for comparing the cost of Diesel automobile engines with gasoline engines.

It was the world's oldest continuously operating automobile company which was to make the Diesel engine practical for automotive use. In 1907 when the primary Diesel patents expired (by specified time limits), engineer Prosper L'Orange, by then with Benz & Cie., used the knowledge now in the public domain to invent and patent a "pre-chamber system" for mechanically injecting liquid into a first-stage combustion chamber. This eliminated the extra tank and made possible the development of a fast-turning engine for functional use in an automobile.

Even this brilliantly conceived idea did not lead immediately to a really effective automobile engine. Actually it came into being nine years after Diesel's death in the form of a two-cylinder tractor engine which developed thirty horsepower at eight hundred revolutions per minute. During the following year (1923), also from the Benz laboratory, there emerged the first four-cylinder Diesel truck engine with fifty horsepower at one thousand revolutions per minute.

In 1927 when Robert Bosch developed an injection mechanism which synchronized the metering of fuel with the speed of
the engine, the place of the Diesel engine in the passenger automobile was finally assured. Even so, the Mercedes-Benz, Model
260D (of 1936), turned out to be the world's first mass produced
passenger automobile with a Diesel engine. Its production was
a brilliant automotive achievement, but it was made twenty-three
years after Diesel's death.

Personal rather than professional reasons may first have caused
Diesel to try to adapt the engine for automotive use. His son-in-
law, Arnold von Schmidt, was an automotive engineer who may
have whetted Diesel's interest.

Even before his daughter's marriage, Diesel had been interested in automobile races. In 1900, the inventor was among the
interested spectators at the start of the very first cross-country
automobile race from Paris to Vienna. He noted appreciatively
that the winning machines traveled faster than the speed-run
records of the better railroad trains—at that time forty to fifty
miles an hour.

During 1904, Diesel attended the start and the finish of the
renowned Gordon Bennett races, whose speedway included the
steep-graded roads across the Taunus Mountains in Germany
near the Belgian frontier. The Kaiser was also among those
present for the race. Driver Jenatzy, who had won the previous
Bennett race—368 miles over less precipitous roads in Ireland
with a 60-horsepower Mercedes racer at an average speed of
55.3 miles an hour—was on hand as the favorite to win again.

The hilly circuit involved 331 miles of cross-country driving.
Its winner turned out to be France's entrant, Jacques Thery,
who drove a Richard-Brasier racing car at a course average of
54.5 miles an hour. Camille Jenatzy and his Mercedes came in
second, but still with a miles-per-hour superiority to any comparable run by a railroad train, even on level land.

Diesel was described as being more excited than he had been
since his own first engine ran. On his return to Munich he bought

his first automobile, a 20–24-horsepower NAG touring car, painted bright red and capable of an average speed of around thirty-five kilometers—twenty-two miles an hour.

But the greatest engine developer of his own or any other century never became a skillful driver. He had handled horses passably well, but the chugging, dust-splattering automobile troubled and fatigued him. It must be remembered that the inventor did not buy his first automobile until he was forty-six years old, by which time his right foot—even then the brake foot—was gouty and his vision not as sharp as in his youth. Then he made a sensible move—he employed a capable chauffeur.

Martha and the children, at least for the time, followed a hands-off policy. So, in the main, did Diesel. When the NAG automobile nagged or declined to function properly, as it did too frequently, the owner rarely tampered with its delicate parts. When it stalled on the road, as it sometimes did, Diesel provided the chauffeur roadside advice on the care and remedy of the balking engine. He rarely went farther, practically never dirtied his hands. There may have been good reason: he was not well, and he believed in letting others do the work they were paid to do.

The inventor continued to use the horse-and-carriage barn which stood behind the Maria-Theresia-Strasse villa, discreetly screened by high privet hedges. He retained the team of handsome claybank carriage horses and the enclosed phaeton trimmed in brown leather with formidable baggage trunks behind. For business and extensive pleasure trips, he continued to travel by train and after arrival by hired rig or limousine.

While Diesel's health trips were usually to the Alps, the family "social vacation tours," inaugurated in the early 1890's, generally included the more fashionable *Kurorte* elsewhere. Diesel had branded these tours a waste of time, but as the decade ended, his view of recreational travel seemed to have softened. It did become expedient, however, to merge business and pleasure. It was with this in mind that Diesel planned what at that point was the longest trip of his much traveled lifetime—his first visit to the United States of America.

The Inventor as Investor

IN JUNE, 1904, Rudolf Diesel set forth on what he insisted was a recreational junket. From the beginning, however, he made it something more than that. At Kiel, where he briefly visited Krupp's Germania shipyards before boarding the liner *Pretoria,* the tall and slightly balding inventor began taking notes.

Although Diesel's fellow passengers on the ship were of varied social, economic, and educational backgrounds, they mingled easily and freely on the voyage. "When the shores of Europe disappear," Diesel wrote, "so somehow, and surely for good reason, do the class distinctions of Europe . . . ." Diesel joined the sociable exchange; he liked people. He was professionally interested in the ship's rather complicated power plant and its miscellaneous power accouterments from bridge to fantail. The westward journey was pleasant and thought-provoking.

On the ninth day, the S. S. *Pretoria* rounded Sandy Hook, docked briefly at immigrant-swarming Ellis Island, and took on a pilot and a small party of newspaper reporters and photographers. Apparently, and quite probably at his own request, the famous inventor was not photographed by Manhattan's working press, though several of his fellow passengers were. His arrival did produce a three-line notice in the "Ships' Arrivals" column of the *New York Morning World,* June 20, 1904, in which he was mentioned simply as "Dr. R. Diesel, the famous inventor from Munich."

Diesel's first impressions of New York apparently were hazy. He spent two nights at the old Waldorf-Astoria Hotel, on the site of what is now the Empire State Building. He briefly visited the Diesel Motors of America at 11 Broadway, where he signed the guest registries. His known letters made little mention of the city other than to note that Baghdad on the Hudson was sultry and noisy, and its streets inconveniently littered with horse dung. He was mildly interested in American styles of electric streetcars and much interested in the hydraulic elevators still in rather common use. He mentioned the painful slowness of the contraptions and the peculiar slurping noises produced when the operators began overhanding the lift rope.

The railroad trip to St. Louis was disappointing. The trains were bulkier and heavier than German trains, but not as clean or as well attended or as fast. His train averaged only about twenty-three miles an hour. He spoke favorably of the food in the diner and noted that the Negro waiters appeared to be the best disciplined and most personable members of the train crew.

At St. Louis both Adolphus Busch and Colonel Meier were awaiting the arrival of their eminent guest. Diesel was doubly warmed—by their hospitality and by the summer sun of St. Louis.

As for the world's fair—the Louisiana Purchase Exposition— he was amazed and disappointed to find so few exhibits of what he termed "technical and scientific achievements." To his alert eyes the St. Louis fair was mostly a country fair made big and just a bit bawdy. He was amazed to see the variety of amusement devices—roller coasters, side shows and carnival tents, cotton-candy dispensers, Ferris wheels, "crazy railroads," and hundreds more. The hordes of market stands seemed to him to mark the exposition as a poorly planned merger of market center and oversize carnival.

But Diesel liked the people and the food. He was pleased to note that Americans seemed to be eating almost continuously. He quickly grasped the reason for this impression. He found the

American food delicious, pork chops and roasting ears of corn—not necessarily mincy-wincy sweet corn—but good, honest, field corn! He made his first acquaintance with American barbecue, including ribs of both pork and beef. He liked both. But he did not especially like American beer; he found the effect of bourbon whisky made from corn "like being struck on the head with a mallet." He thought American girls quite pretty but not very tastefully dressed. He particularly liked American farmers and livestock men.

The inventor was not lionized on his first visit to St. Louis. That he was not could have resulted from the superabundance of famous visitors and from the fact that the St. Louis newspapers —there were at least six—were overwhelmed with "fair news" and understandably were crowded for space. But then Diesel had come to learn, not to talk.

Then, rather abruptly the gracious, slightly limping visitor took leave of the Busch mansion and the "inhabited mausoleums" nearby and boarded a westbound train. He noted regretfully that he could see no extensive American demand for the Diesel engine, however efficient, long-lived, and economical in its use of fuel. This amazing "Western world of a nation," as the inventor dubbed it, had oil and coal to waste and was wasting them in grandiose fashion. American fuels were incredibly abundant and, by contrast to European, quite inexpensive. Yet American industry, youthful giant that it was, continued to be almost viciously competitive and unbelievably cost conscious. Diesel engines were superior but expensive; in this respect there seemed no likelihood of change.

Rudolf Diesel wanted to learn more of this wonderland. At Denver he forsook train travel temporarily, hired an auto, and made an impromptu tour of the Pikes Peak area, where he was deeply impressed by the Garden of the Gods.

Returning to Denver, he entrained for San Francisco. The inventor liked San Francisco enormously and described it as one of the truly charming cities of man. The City of the Golden Gate presented an alluring merger of East and West; it was a tumbling together of Russia, China, and an American frontier which had gone bounding west to meet the Pacific. The "melting pot" (this cliché both amused and annoyed him) had not yet fulfilled its name, but it was boiling lustily and, on the whole, gaily.

Eastbound again, Diesel made a stop to visit the Mormon country. He liked the Latter-day Saints and found Salt Lake City one of the most remarkable settlements he had ever seen. He could not believe the place real; even so, he liked it.

Diesel returned to Munich without further side trips. Apparently he had found the American journey salubrious. For almost a year he did not complain of headaches, gout, or other ailments. If he suffered pains, they were evidently less bothersome than those preceding his adventure and, alas, than those which would presently occur.

Apparently Diesel spent much time pondering the promising future and unimpressive present of the American Diesel Engine Company. The Busch-Meier team was "carrying the banner," but thus far it was neither making nor selling Diesel engines. The inventor had very real affection for both Busch and Meier and was certain he desired and needed their friendship, but at this point there was no substitute for the actual sale of his product.

Rudolf Diesel needed money. He was, as he had heard the Americans say, going broke hell for leather. As he freely admitted, he had not developed a sales resistance comparable to his own skill as a salesman. More regrettably, he had repeatedly dipped into investment areas with which he was not familiar and which he had not bothered to investigate.

Apparently without benefit of experienced counsel, he had invested heavily, no fewer than half a million marks, in Galician oil promotions. One after another the test drillings had failed to deliver the black gold. Frenzied but belated investigations indicated that the future prospects of striking oil were poor or non-existent. The promotional tricks to which the inventor had been subjected fringed on the poorly defined boundaries between dubious legitimacy and criminal fraud. Diesel was a distressed victim; to the cynical or less than sympathetic, he had been "played for a sucker."

By the time of the inventor's return from St. Louis, his real estate investments, also made with the spur-of-the-moment incompetence of a romantic schoolboy, totaled approximately one million marks. The current value of the properties, including proposed sites for factories, reservoirs, and suburban developments which never materialized, was probably no more than half the total invested. The real estate quandary was destined to become still worse before getting better—much worse and only slightly better.

His largest single investment was in the "directional pool" of *Allgemeine Gesellschaft für Dieselmotoren,* two million shares with problematical values. Some guessed one figure; others, another; but since the stock never paid any dividends and was never listed on any accredited exchange, the estimates were of no real consequence. *Allgemeine* was an impossible alliance and a lost cause.

Similarly, *Dieselmotorenfabrik* Augsburg A.G., founded in 1898 to manufacture Diesel engines, was never a profitable undertaking. A firm formed specifically to build the engine which other firms seemed too slow to exploit, the newer *Maschinenfabrik* never functioned practically. The causes are obscure, but the venture failed.

For years the directors, with an eye for protective finance— that is, protecting themselves from bankruptcy suits—had urged Diesel to "absorb" the outstanding stock. Persuasive bankers had joined in the chorus. Their reasoning, to use that word partisanly, was that if the factory should collapse in bankruptcy, the prestige of the engine would be mired, its escutcheons would be blotched, and Diesel power development would suffer globally.

In retrospect their arguments seem smug and ridiculous. The preponderant and perennial causes of bankruptcy are inept management and marketing, rather than failure of the product or service dispensed. But Diesel, with an artist's pride in his invention, yielded to persuasion and purchased all the outstanding stock of the corporation. In 1911, after long and unproductive dormancy, the corporation was liquidated, with still another disastrous loss to Rudolf Diesel.

The inventor's financial nightmare continued. His increasingly frantic trips to the Galician oil fields proved repetitiously unsuccessful. He not only failed to discover tangible assets but found himself frustrated by his inability to negotiate with Russian authorities for assurances of protection for future financial investments or for restitution.

By 1906 he found himself divorced from what had been his best manufacturer, the M.A.N. plant at Augsburg. After involved litigation, the inventor won a court decision the following year, but for several years to come he was alienated from his best-proved homeland production base.

It is estimated that by 1905 his losses amounted to between three and three and one-half million marks. He saw his savings melting and his once substantial credit rating falling in shambles. It is not surprising that the headaches and swelling of the feet began to recur.

Diesel was determined to recoup at least part of his formidable losses. On the second floor of his mansion on Maria-Theresia-Strasse, he resumed his work to devise a way for building more and better engines for ships, locomotives, and automobiles as well as for stationary power plants, pumps, and pipelines. Once more the creativeness of mechanical design reigned in his renewed infatuation with work. His one expression of regret was, "It all takes so much time . . . ." Also so much money. He felt his emptying hourglass draining away his chances to regain his lost wealth.

Competing engine makers continued to blast his engine as being impractical. His Berlin lecture dealing with his rational prime mover (the "Charlottenburg Speech") was openly attacked at the time of delivery. The especially promising preliminaries for developing locomotive engines were dragging. Progress with small engines for use in automobiles seemed completely halted. But Diesel was not halted. As opportunity and invitations allowed, he continued to lecture before sociological groups as well as before technical gatherings. Without appreciable success, he tried again to revive his movement for "solidarism"—his plan for workers' sharing the ownership of production facilities.

Renewed activity in his work was matched by renewed interest in other things. He did not return to playing the piano, but during the 1905–1906 season his enthusiasm for opera returned. Munich, fortunately, was maintaining the luster of a capital for the performing arts. Its National Theater reopened to the new work of Richard Strauss, *Salome,* which first-night audiences branded as sensuous, shocking, unprincipled, and totally delightful. Diesel agreed only with the last adjective. He found in the "new" operas a sense of spiritual resurrection.

In addition, Diesel tried to revise his principal family recreation—vacation trips. Previously, such trips had been hardly more than rituals of status, a fault for which he and his wife shared joint responsibility.

Consider, for example, the time the five Diesels had vacationed in style in Italy in 1899. The family started off by taking an express train, first class naturally, to Trieste. There they lodged at the Hotel Trento. Thence, they rode by a hired landau drawn by four horses (two white and two brown) to the elite resort town Madonna di Campiglio. The heavier baggage was shipped on ahead. Having first been wrapped protectively in linen covers, the hand luggage (for other hands to lift, of course) was stowed on the back rack of the elegant carriage. From the Grand Hotel the entire family in de luxe walking toggery hiked to Monte Spinale to view the superb scenery.

A happy change was the summer outing of 1907, which was recounted by the younger son, Eugen, in his book, *Jahrhundertwende. Gesehen im Schicksal meines Vaters.*

The inventor, then nearly fifty and with graying hair, loaded his wife, daughter, and two sons into the fiery red family touring car with the round bronze radiator and boldly set forth for Paris with himself at the wheel. Their eventual destination was the *Conservatoire des Arts et Métiers*, the solemn and fortress-like museum shrine of Diesel's childhood. The special mission was to pay tribute to Denis Papin, the physician and inventor who, in 1690, had given the world *la machine à vapeur*, in principle the undying steam engine.

The first stage of the pilgrimage included spending an entire week in Blois on the Loire, where Papin was born and had grown to manhood. Diesel was fully convinced that in order to know any man, one must first gain acquaintance with his formative surroundings and environment. He admired Papin deeply, explaining that D. Papin, like R. Diesel, had been "born too early for his work." From languid Blois, the expedition chugged on to Paris and the *Conservatoire*, where a heroic statue of Papin stood in the entrance yard to the one-time cloister. There the Diesels stood in silent homage.

The family returned to Munich barely in time to attend the funeral of the inventor's father. Following nearly twenty harmless and happy years of stargazing, magnet fondling, and conducting séances attended principally by himself, the elder Diesel passed away quietly into the grand séance called eternity. His widow, Elise, chose to live by herself, occupying her time by visiting her widowed daughter Emma and claiming more attention from her always preoccupied son.

The illustrious, busy, and direly harassed son responded to her wishes; indeed, he showed a renewed warmth and affection in other family ties. His appearances in company of his wife became more than occasional. Although his elder son remained aloof, apparently much lost in private plans and dreams of being an artist, the inventor appeared to grow closer to his younger son Eugen, who quietly adored his father and expressed every intention of following in his footsteps, beginning with factory service with the Sulzers of Winterthur.

Daughter Hedy, since birth the apple of her father's eye (the metaphor is also old Bavarian), was now at that wonderful convergence of affections which blesses young womanhood. She was nurturing both the love of her father and that of a younger man: his name, Arnold von Schmidt; residence, Frankfurt; intentions, marriage, making automobiles, building a home, and begetting children. Diesel liked and respected his prospective son-in-law but with that ever baffling masculine paradox—a father's resentment of a younger man who would take away his daughter.

Conversationally Diesel was quite willing to grant that a girl child is born to be loved and comforted, to grow up, marry, and bear children; also that her "little girl's" love of father must flower and fruit in her more consuming love of mate . . . . But, again, Rudolf Diesel preferred to finish the statement on another day.

With rather grand nonchalance, therefore, the inventor planned one more vacation, with his daughter as his only and very special companion and guest. It may well have been the most enjoyable vacation trip of his life.

Hedy's wedding to Freiherr Arnold von Schmidt was among Munich's outstanding social events in the season of 1908. The groom and his family arrived punctually from Frankfurt-am-Main for the formal Lutheran service. Although wracked by headaches, Diesel endured the prenuptial parties and dutifully gave away his beloved only daughter. He appeared briefly at the lawn-party reception afterwards and took his place in the receiving line. He exchanged banter with the bride and groom, smiling wanly while dubbing his new son-in-law a "pretty-daughter stealer."

Arnold accepted the label, pointing out that however homely and trifling a groom is, he is nonetheless necessary for a wedding. As Diesel freely admitted, von Schmidt was neither homely nor trifling; indeed, he was quite personable and brilliant. He had already earned a degree with honors in mechanical engineering from Diesel's changed and expanded alma mater. He was profitably employed as an automotive engineer by the Arnst automotive works of Frankfurt and was on his way to becoming a factory manager. "We won't leave you to wait alone, dear father," Hedy consoled. "We'll be bringing you grandchildren, beautiful grandchildren." Hedy was a young woman who kept her word.

After Diesel bade the couple farewell, he abruptly returned to his quarters and went to bed. Next day he complained of an excruciating headache. The following day he quickly arose, packed his valise, and left on a business trip to Vienna. While in the city of waltzes he began work on the last and most serious of his books, *The Origin of the Diesel Engine (Die Entstehung des Dieselmotors).*

After the inventor's return to Munich, he plunged again into a painful and confused series of attempts to relieve his now nearly desperate financial straits. He was practically penniless, shorn of credit, and somewhere near half a million marks in debt. Apparently Diesel was unwilling to admit his growing multitude of financial indiscretions, even to himself. Instead of seeking out bona fide bankers and respected investment brokers (his list of personal friends included several of each), he sought no counsel and accepted no advice. Instead, he plunged into "deals" which were not prudent and by some standards were distinctly lunatic. One of these involved the purchase of a rather dubious public lottery. The sellers happily pocketed the inventor's cash without any contractual promise not to set up a rival lottery immediately. Indeed, the cheaters promptly did just this, whereupon Diesel again lost every penny of his frantic investment and ended the wretched play still deeper in debt.

Without any substantial improvements of his headaches, gout, or financial worries, Rudolf Diesel collided on all fronts with the inevitable quandary of expiring patents. As one after another "died for time," he was obliged to obtain renewals and renegotiate contracts with manufacturers.

In 1910, the inventor found it necessary to visit St. Petersburg to establish a new agreement with Emanuel Nobel. He accomplished that with a marked degree of success. Martha accompanied him on the long and trying Russian journey and proved a helpful companion. Emanuel Nobel received his callers graciously. He was quite certain, as Diesel reported his words, "that Russia was already pulling herself out of the ice floes . . . ."

Excepting only Germany, Russia was then using more Diesel power than any other nation. Nobel was eager to extend his franchises and to expand both the manufacture and the placement of the petroleum-consuming engines. As need required, he was willing to import the engines from German and Swiss sources.

The presence of Diesel's wife was not in keeping with prevailing business protocol of the Russians, but it indicated the return of some of Martha's long-waning wifely interest in her husband's business. Her presence in St. Petersburg also contradicted to some extent the gossip that Diesel's wife was no longer his helpmate.

Consider, as many then were doing, the admirable wifely example of Bertha Benz, who was still helping husband Karl Benz blaze his way as Germany's and the world's first great automobile builder. Bertha had encouraged and "babied" Karl to keep him at his work during his darkest days and years. Valiantly she had served as her husband's first test driver, at a time when test-driving an automobile was more laborious than and fully as perilous as test flying a "poultry crate" airplane was later to be. Bertha Benz was all woman, and all wife. But by way of contrast, take that Martha Diesel . . . .

Rudolf did indeed "take that Martha Diesel" to Russia, for what may well have been their most portentous journey together. It prepared the way for a man-and-wife journey to America.

As a result of his talks with Nobel, Diesel grew excited about the opportunity to speed up acceptance of his engine in America. The American franchise continued to lag in terms of the actual number of engines manufactured. Busch granted the point, repeated his intentions of "getting into production in a big way," and willingly renewed the inventor's franchise. For good and generous measure the beer magnate presented to Diesel a number of shares (some $150,000 worth) in the reorganized American company.

Diesel had been able to align the Sulzers of Switzerland as mutually beneficial allies of Busch. The Winterthur pioneers had supplied the American Diesel group with model engines as well as with sales pointers and prestige. The American Diesel Company was shortly to be reorganized as the Busch–Sulzer Brothers Diesel Engine Company. Plans were already drawn for building an American manufacturing plant in St. Louis. The inventor eagerly agreed to serve as honoree for the ground-breaking ceremony. He remained quite certain that America would be the promised land for his engine.

Diesel, thinner than ever and rapidly graying, carried on his engine research. Although increasingly haggard and in continual pain, he sought to keep up what he termed his "social sweepings." He accepted an invitation to serve as a judge of the mechanical exhibits of the Turin (Italy) world's fair. In 1911 he represented German technology at London's World Congress of Mechanical Engineers.

There, at the "grand banquet," Diesel took a place of honor beside Sir Charles Algernon Parsons, who in 1884 had invented the steam turbine. After paying warm tribute to Parsons, Diesel delivered a "technical address" in English and without recourse to the first person. He was enthusiastically applauded by the international assembly of renowned engineers and industrial leaders.

Returning to Munich, Diesel set to work completing plans for large marine engines, for the most impressive engine progress was then concerned with ships. Diesel power had carried Roald Amundsen to the discovery of the South Pole (December 14, 1911). It was helping power the magnificent liner *Christian X* of the Hamburg-American line. The Imperial German warship *Prinzregent Luitpold* was to be powered partly by the M.A.N. works at Augsburg. Already completed was Howaldt's motorship *Monte Penedo,* powered entirely by Diesel engines built by the Sulzer Brothers. In schools throughout the world, pupils and teachers were obliged to stand ready to revise the conventional drawing-book version of the steamship with stacks spurting black smoke. For the coal-stoked ship, the end was in sight.

Herr and Frau Diesel Go to St. Louis

꜕

꜕

꜕ LONG BEFORE THE OMINOUS YEAR 1912 began, Rudolf Diesel had resolved to visit America again. The engine had been born in Germany and passed its early childhood in Europe. The inventor never doubted it would reach maturity in America. He felt both duty bound and privileged to be more in step with this wonderland to the west.

Diesel knew that the reorganized American company needed his presence, probably for counsel and certainly for prestige and publicity. Both Adolphus Busch and Colonel Meier had stressed the latter. The inventor recognized the situation and accepted the necessity of appearing as a public figure. On his first journey, he had kept mostly to the role of solitary observer and thoughtful note-taker; this time he would have to be exhibited, even if as an ailing or, just possibly, a failing lion. A lion, even if ailing or failing, requires a keeper or a companion.

Prior to the year's beginning, Diesel had decided to take his wife along. Quite probably Martha had done part of the deciding. In any case, on the trying and fatiguing trip to Russia two years before, he had enjoyed her companionship. The inventor could no longer deny that, at least on a long and fatiguing journey, a good companion was helpful. He was not a well man. His ruinous headaches still recurred. His gout was no better; like the headaches it plagued Diesel with intense and repeated flare-ups.

Rudolf and Martha set about making arrangements with care.

This, of course, included the choice of an appropriate passenger liner. The British White Star Line had a fine new super liner, the S.S. *Titanic*. The Diesels would have liked to book passage on the maiden voyage of the wonderful and "unsinkable" $7,500,000 *Titanic*. As one might have expected, a great many others, including many wealthy Americans, had the same idea.

By March of 1912, the Diesels learned that the maiden voyage of the *Titanic* was scheduled for mid-April. This was too late. Busch had already approved plans for formal ground-breaking for the new American Diesel engine factory on April 14, with the inventor lifting the first spadeful of earth.

Furthermore, Diesel would be speaking—lecturing—at American universities. It was important that he appear while the regular terms or semesters were in progress. Moreover, most of the important engineering societies held their annual meetings or conventions in late April or May. It would be better, therefore, for Herrn and Frau Diesel to arrive in New York late in March.

American interest in engines was rising. At the time of Diesel's solitary American journey of exploration eight years earlier, the United States as a whole was not yet "engine minded." The horse-and-buggy era had only recently passed its peak of some thirty million units a year. In 1904 the American automobile industry was lagging far behind Europe's; it was still speculative, imitative, and uncentered. The automobile "capital" was then in Springfield, Massachusetts, but later was lured westward to Indianapolis and South Bend, Indiana. At that time, when even one-automobile families were few in number, especially as compared with two-buggy families, Detroit was of little consequence.

But by 1912 the American automobile era was beginning to gain momentum. Every working day was stretching its lead. Although the Diesel engine was not yet adapted for the automobile, many mechanics and designers already believed it would be.

American consumption of all industrial power was catapulting. Since 1910, hydro-electric plants and public water supplies had been responsible for the greatest increase in horsepower. Hundreds of municipal power or public water plants were being renovated or built. For use in these generators of power, the Diesel engine held special promise.

American newspaper coverage of the second Diesel visit displays an awareness of its news-making importance largely wanting during the inventor's 1904 visit. Adolphus Busch, his men, and his industries were unquestionably publicity conscious: the American Diesel company had employed a press agent. The Busch-men sensed the background of what was already being called "mass interest."

Those who remained of the New York office staff of the American Diesel Engine Company (the Busch-Sulzers reorganization was principally in St. Louis) were at the pier to greet the distinguished visitors. The secretary-general of the American Society of Mechanical Engineers was also present; the directors of the Society had already voted to make Diesel an honorary member.

As soon as they landed, the Diesels found themselves rated as "live" or "hot" copy. Flashbulbs popped in their faces. Wherever they went, the couple was besieged by newspaper reporters—in hotel lobbies, lecture halls, theater lobbies, even on the sidewalks, on their morning strolls around Gramercy Park.

Both separately and together, the Diesels were at times almost buried under reportorial questions. The newspaper accounts provide a basis for two general deductions: New York newspapers of 1912 were dull, tactless, and in the main rather badly written, and both the Diesels were determinedly gracious and discreet. When newswomen asked for the views of the inventor's wife on woman's suffrage, her answer was, "It is an interesting development." Asked her views on American political figures, including Teddy Roosevelt, then in the throes of splitting the long-dominant Republican party in another bid for the Presidency, she answered, "He is an interesting development." What was her view of the growth of the Lutheran church in America? "Lutheranism is always an interesting development."

When and as permitted, Diesel spoke of the place of his engine on the American scene, present and future. He dwelled perhaps prematurely on the perils of air pollution. As he had once explained to his children when they seemed distressed by the occasional puffs of exhaust smoke from the central chimney of the Diesel engine pavilion at the Munich exposition, any working engine emits some amount of exhaust waste, but the Diesel engine was the cleanest running of all.

Steam engine operation—as anybody could see with, or alas, feel in his own eyes—meant foul-smelling, smogging clouds of dark smoke and unending rain of unburned particles of coal dust. But not the Diesel engine. It could, would, and did consume any kind of fuel, from butter to bituminous coal, cleanly and efficiently. Consider, my friends, the new Diesel-powered oceanliner which had no need of smokestacks, or those ships still using smokestacks, yet showing only slight gray flurries of exhaust while moving ahead of the old-style smoke-cloud raisers.

Already that American wizard of inventions, Thomas Alva Edison, had publicly stated that the development and improvement of the Diesel engine (in 1911) was one of the truly memorable accomplishments of mankind. The world's total of Diesel power was already crowding one and three-quarters million horsepower and would reach the two million mark by the end of 1912 (with only a little more than one-third in the homeland, Germany).

The press recorded what the famed inventor said, and even more he did not say. Quite probably the biggest feature was in the *New York American* of Sunday, April 14, 1912, under the ink-dripping headline: DR. DIESEL EXPLAINS HIS ECONOMICAL CRUDE OIL ENGINE.

When Diesel was quoted by the *New York World* (April 2, 1912) as saying that his engine was the "most simple and natural of prime movers—not just another improvement on an older type of internal combustion engine," he was striking fire in the working American language. When he reiterated, "The Diesel is the only engine which converts the heat of natural fuel into work and that in the cylinder itself without any previous transforming process . . . . It has become for all liquid fuels what the steam engine is for coal, but in a much simpler and more economical way," anybody could see that the inventor of the rational heat engine was also its rational interpreter.

As a group, the technical periodicals had even better cause for seeking out the visiting inventor. Mirrlees of Scotland had lately built an impressive new engine factory near Stockport in Cheshire. The goal of the new enterprise was to perfect and build bigger and better Diesel power plants, up to 750 horsepower. Now, in 1912, Mirrlees was making the first Diesel electric installations on big ships. It was also marketing an enclosed engine to develop eighty horsepower per cylinder at a moderate speed—250 revolutions per minute. Also for the first time, Mirrlees was successfully making and selling multi-engine installations especially designed for operating power plants.

Mirrlees was on the verge of two other innovations destined to make history. One was a special type of Diesel engine suitable for powering the battle tank, which was to be an important factor in the Allied victory in World War I. The second proved to be the salvation of the soon to be beleaguered British Isles. This was a pilot injector which enabled the Diesel engine to run successfully on a fuel mixture of up to 90 per cent of coal-tar oil.

Werkspoor of Holland, on order of Royal Shell Petroleum, had perfected and put in use a directly reversible Diesel engine for propelling ships. The enclosed four-cylinder engine was built with a cast-iron frame; the engine was water cooled and pressure lubricated. Greatly admired by engineers, the "improvement" had been honored at the Brussels exposition and was licensed for manufacture in Germany, France, Great Britain, and the United States.

There were also advancements in Italy. Franco Tosi of Legnano began manufacturing in 1907 with a single-cylinder, 30-horsepower, upright unit. By 1912, Tosi had launched into design and construction of big engines for big power plants; a 2,400-horsepower Tosi was soon, though only briefly, to be the world's most powerful single-unit Diesel engine.

Meanwhile, the irrepressible Sulzers of Switzerland had begun building king-size engines. In 1910 they built a single-cylinder engine with a 1,000-millimeter (39.37 inch) bore which developed 2,000 horsepower at 150 revolutions per minute. The successor, begun the following year, was a four-cylinder "monster," each cylinder developing 850 horsepower. The four-cylinder, 1,600-horsepower "standard" was already on the Sulzers' drawing boards.

As the inventor knew, for he himself had spent the first two months of the year working on it, a Diesel engine for locomotives was also in process. The pilot model which Diesel had especially encouraged was a four-cylinder, two-stroke-cycle, V-type engine rated at one thousand horsepower. It acted through coupling rods placed directly on the driving axles. The first model was driven by air compressed and stored in large air receivers by an auxiliary Diesel engine. This engine had not yet been put to practical work. The forthcoming tests on the Prussian State Railways were to show serious faults, such as difficult ignition due to the cooling of the compressed air, and inadequate overload capacities at low speed. Indeed, Diesel's first concept of a locomotive engine never worked effectively until an electric transmission system was installed; the first successful user was the Prussian and Saxon State Railways in 1914.

In his first American platform appearance, Diesel stated frankly: "The inventor does not wish to predict whether or not his attempt at an entire revolution in the working of railroads will be successful at once . . . . But one thing is certain—the Diesel locomotive will come sooner or later . . . ."

The same statement, in slightly varying words, was to be included in his three other principal American addresses—at the United States Naval Academy, Cornell University, and the convention of the American Society of Mechanical Engineers at Cooper Union in New York. In each of his three most-heeded

American lectures, Diesel asserted that even though most Diesel engines were then in stationary work, his engine would have its greatest value in transportation. "Nowhere in the world are the possibilities of the prime mover as great as in the United States of America," he declared. The inventor's foremost interest at that time was the development of a workable locomotive powered by his engine.

Of all his lectures so painstakingly prepared for American delivery, none was so burningly sincere as the least-noted one, which he delivered in St. Louis's Old Mechanics Hall under auspices of Herrn Adolphus Busch. The date was the evening of April 13, chill and dreary. The host audience included five hundred members and guests of the Engineering Society of St. Louis. Diesel appeared half-an-hour early, impressive in swallow-tail coat and white tie. He endured stoically the somewhat flatulent introduction and bowed briefly and stiffly to the first scattering of applause. He spoke in crisp English, virtually without accent. As usual, he meticulously avoided the first person.

He began by explaining that the Diesel engine was a "new method of the internal working process, not a structural improvement of older engines." He recalled the feature exhibit at the Turin exposition of the previous year, a "work team" consisting of a typical steam turbine and a crude-petroleum-fueled Diesel engine, built and exhibited by Tosi of Legnano.

"The difference between the two plants was this," Diesel summarized with an overlength sentence: "for the working of the steam engine the complete boiler plant with its chimney, fuel supply apparatus, purification plant for filtering the water, with feed pump and an enormous quantity of water had to be provided, with the final result of consuming two and a half or more times the fuel per horsepower as was required by the Diesel engine standing beside it. The latter being an entirely self-contained engine without any auxiliary plant, took up its crude fuel automatically and consumed it direct into its cylinder without residue waste."

The inventor commented on power fuels. He recalled that

in 1899 he had developed an experimental engine which he successfully fueled with coal tar and creosote oil. He stated his belief that coal "fractioning" would soon be developed to a point that "in case of war and the consequent cutting off of the supply of foreign fuel, production would be sufficient for running the entire fleet, both war and mercantile, and for providing in the meantime the power necessary for land industries."

Diesel repeated that his engine was a proved means of breaking any fuel monopoly—"not by law or other artificial means but by the invincible force of scientific investigation and industrial progress." He summarized recent progress in equipping his engines with atomizers which permitted the use of castor oil, palm oil, lard, and various other "natural fuels." He recalled that he had exhibited at the Paris exposition of 1900 an engine which ran on the oil of the African earth nut *Arachis hypogaea*, a "plant grown in the tropical wilderness." He added: "The use of vegetable oils for engine fuels may seem insignificant today. But such oils may become in the course of time as important as petroleum and the coal tar products of the present time."

His next thesis was that engine fuel is inevitably the decider of the fate of nations, that the ability of an engine to consume any flammable surplus locally available decides basic economies. He inserted prophetically: "Motive power can still be produced from the heat of the sun, always available, even when the natural stores of solid and liquid fuels are completely exhausted."

Then the philosophical thermal engineer changed back to a reminiscent engine builder. He described the first vertical engine which "the author" had envisioned and assembled in 1893 (when he was a "mere" thirty-five. He described it as a "steel cylinder filled with a piston rod, also with an external crosshead, the cylinder having no water jacket . . . . The starting storage chamber consisted of a wrought-iron pipe with riveted flanges; there was no air supply pump, the fuel being injected directly.

"The engine never succeeded in running, not even one revolution, because at the first injection of fuel . . . there occurred a terrible explosion . . . nearly killing the Author. But the Author knew then just what he wanted to know . . . . It was proved possible to compress pure air so high that the fuel injected into it ignited and burned."

Thus the Diesel engine had been born—dangerously. This might-have-been-fatal explosion, however, verified the Diesel concept of a rational heat engine as basically a tightly enclosed pipe inside which a piston is used to compress the air until enough heat is generated that the fuel will ignite in "the red-hot air."

The first engine model, fueled by powdered coal, had exploded. The second, designed to burn gasoline, also proved unworkable; it never ran more than two strokes at a "firing." As Diesel recalled, "It was always dangerous to stand near. . . . But it gave the Author the first indicator cards of the whole cycle [of internal combustion]. These first two engines taken together proved the possibility for carrying out the combustion process which the Author had recognized as possible years before. . . . The concept was regarded as impossible by the technical world of that time."

The inventor repeated that his initial conviction had not been based on pipe dreams or poems but on "infallible mathematics." He recalled that almost exactly fifteen years before the evening of this lecture and after four years of "laborious experimenting," he had produced in his Augsburg workshop an engine which proved to be commercially practical. It was an upright or vertical stationary engine, standing fifty-seven inches high and weighing about six hundred pounds, including a starting flywheel about as big as an ordinary wagon wheel. This was the model which Colonel E. D. Meier had originally appraised and "sold" to Busch.

Diesel shifted skillfully to issues of basic economy: "Nowhere in the world are the possibilities of the prime mover as great as they are here in the United States of America!" Then he warned, "The Diesel engine is not and will never be a cheap engine. It aims to be the *best* engine and must be constructed with the very highest class of material, with the best tools, and by the most skilled workmen . . . ."

"The leading idea in Europe is always the economy in operating cost; the leading idea in American economy is the *first* cost. The word 'efficiency,' which is the base of every contract in Europe, seems to be unknown to a vast proportion in this country—not engineers but to business men and to purchasers of engines."

Diesel granted that steam engines and gasoline engines were "much cheaper to operate" in America than in Europe, because America was so "monstrously" rich in both coal and petroleum that it could and did afford "gigantic wastes in power production." But he insisted that economy in power fuel must inevitably transcend the whimsey of any nation or the God-given abundance of natural resources, even in this "country of mighty oil wells and gigantic railroads."

With the deft casualness of an expert bridge player holding a handful of tricks, the inventor shifted to marine engines. He said that he had sought to develop ship engines not as a marine engineer but rather as a "landlocked thermal engineer." The basic challenge had been that of changing his upright or vertical stationary engine to a horizontal or "lie down" model capable of transmitting greater power at higher speeds to ship or boat propellers and eventually, by due process of gears, to land vehicles.

The first horizontal Diesels were small engines with thin walls, reversing gears, and extremely hard, brazed-steel castings. Almost magically, from the beginning they had succeeded.

His marine model had been developed in France, principally in Paris, by two able French engineers, Adrien Bodet and Frédéric Dyckhoff. The merging of talents had developed a single cylinder housing two pistons which worked in opposite directions, thereby establishing a four-stroke ignition cycle. This permitted a single "power tube" operating at high speed and with good balance to drive a canal boat or other light craft with exceptional competence and decided economy.

Within three years after the first trials on a Belgian-owned canal boat (in 1902), the Diesel marine engine was powering French submarines as well as diverse surface craft, including the "Swedish Navy's submersible gunboat," the *Hvalon*. The latter "wonder craft," powered by three horizontal Diesel engines, had lately traveled underwater for seven hundred knots and "surface skimmed" four thousand knots—all without refueling.

The inventor mentioned the "Dieselization" of the icebreaker *North Pole* as a special example of space economy. The icebreaker had a displacement of 380 tons, at least 100 of them for coal storage. Installation of Diesel power and liquid fuel tanks had reduced fuel space by 85 per cent, fuel weight by 80 per cent, and had increased the cruising range by some three thousand miles.

By 1912, Diesel marine engines were powering a known total of 365 ships, including at least 60 heavy merchant vessels. The list included France's world-leading submarine fleet and the newest and biggest sailing ship. Although the latter used Diesel engines only as auxiliary power, it seemed to intrigue Diesel more than any craft afloat. *La France,* with 10,730 tons displacement, 70,000 square feet of sail aloft, and two 1,800-horsepower Diesel engines below, had been launched November 16, 1911, for speed runs between commercial ports of France and New Caledonia. She was already rated as the most profitable freight carrier of the era.

On the river Amur in far Siberia, the Diesel-powered Russian gunboat *Schtorm* was performing patrol duty on the earth's coldest water run (thermometer readings in upper Siberia are frequently lower than those at the North Pole). On the relatively tropical Caspian Sea, the Czar's gunboats *Kars* and *Adagon* were being powered with the same type of two-cylinder, four-stroke, 1,000-horsepower Diesel units working twin propellers. In both hemispheres Diesel's rational heat engines were powering ships and, as the inventor noted, "multiplying their radius of action by four, cutting engine-crew labor to a fourth, fuel costs to a third, fuel weight to a fifth, and engine room personnel to a third."

The St. Louis lecture revealed that Diesel regarded the success of his rational heat engine at sea as symbolic of its success for driving land vehicles. He dealt lightly with the stationary engines, mentioning only distinguished examples such as the 70-ton battery of "verticals" which were providing electricity for the city of Kiev in Russia, including power for its street railways, and the 80-horsepower uprights which, far down in the hot Belgian Congo, were already powering Africa's first petroleum pipeline.

As a matter of record, more than 95 per cent of all Diesel engines then (in 1912) at work were stationary models. Yet the inventor gave the most time and detail to his newest and, as he saw it, "climaxing invention," the Diesel "thermo-locomotive."

Although the thermo-locomotive had not yet pulled a train or indeed, traveled more than a few yards on its own power, the inventor was certain it would "revolutionize" railroading throughout the world.

It is difficult to imagine a prophecy more audacious—as of 1912—and impossible to name one more literally correct.

Diesel noted that at this point the "thermo-locomotive" had cost him five of his best years. He did not add that it had at that point cost Sulzers, Krupp, and the Prussian State Railways and affiliates at least two million marks and that most of the investors were already writing it off as an eighty-ton black elephant.

"Five years is a long time," Diesel admitted. "But in terms of modern engine building the thermo-locomotive is the most difficult construction problem yet met . . . . Compared with it, the development of the marine motor has been very simple."

With that the lecturer began to demonstrate his remarkable skill at describing a complex machine which not one of his listeners had ever seen and, as it turned out, never would. "The whole plant is contained in a closed engine room that makes the locomotive look from the outside like one of your steel cars. The carriage is 16.6 meters long over the buffers and has two trucks of two axles each, and two driving wheels. The latter are not directly coupled with the engine, but indirectly with the blind axle which is at the same time the crankshaft to the central Diesel engine.

"This engine is an ordinary two-stroke cycle with four cylinders coupled in pairs at an angle of 90 degrees which drive the blind axle whose cranks form an angle of 180 degrees. This disposition gives . . . balancing of the moving masses, the first and most important condition when putting such engines on a movable platform. Between the working cylinders are placed the scavenging pumps which are driven by levers from the connecting rod. Beyond the engine in the roof of the car is placed the silencer. Also under the roof are the canals which lead fresh air to the suction pipes of the different motors and pump cylinders.

"On one side of the main engine stands an auxiliary engine. This latter consists of two vertical-stroke-cycle cylinders coupled to the horizontal air pumps . . . . The air pumps serve according to a special patented process to increase the power of the main engine when starting, also when maneuvering, and going up hill in such a way that auxiliary compressed air and auxiliary oil fuel are conducted into the main cylinder [when needed]. By this means the power is increased, making the engine as elastic as the steam engine."

The audience, after listening intently through the two-hour lecture, got to its feet applauding. Diesel smiled for the first time, bowed, and then headed for the side exit. Waiting was Busch's newest Benz with engine warm and chauffeur standing.

CHAPTER EIGHTEEN

Side Trip: A Remembrance by Charles Morrow Wilson

IT HAPPENED ON OR ABOUT APRIL 5, 1912. The place was Fayetteville, Arkansas, in the Ozarks, on the Frisco Railroad (St. Louis–San Francisco), 356 rail miles southwest of St. Louis.

Word had come from the Frisco telegraphers and dispatchers that an important person would be arriving in Fayetteville the following day. As not many important people ever deliberately arrived in Fayetteville, Arkansas, the local station agent picked up the pipe leak, as the metaphor then was, and passed it along to Leon Smith (local pronunciation Linn Smiff), the editor-publisher of the weekly newspaper, the *Arkansas Sentinel*.

I doubt that Mr. Smith knew much if anything about Rudolf Diesel. As a dedicated publisher, he respected the fact that names are news and the obligation to tend news when it swoops down into one's circulation area.

Fayetteville, Arkansas, at the time was a railroad terminal. From its grimy, red-brick depot and track yard—which imperiled, and still does, the main east-west street, Dickson—two branch lines then operated. One was the Frisco's Muskogee branch, which went west into the Oklahoma foothills. The other was the picturesque St. Paul branch, winding eastward to the timbered backwoods which gave the world wagon-wheel spokes, railroad crossties, red-oak whisky barrels, white-oak beer kegs, and Orval Faubus. Both branch lines had been used experimentally as courses for a self-powered passenger coach driven by two gasoline engines mounted at the head of the unit.

The self-powered coach was called the McKeen car. It had been designed and built by a Union Pacific power engineer, William McKeen, patented in 1907, and first placed in service in 1909. It was the rolling-stock grandfather of the self-powered (usually Diesel-electric) Budd car.

The Frisco Railroad had placed the McKeen car in experimental service in 1910 as a passenger carrier on short branches where traffic did not justify operating a locomotive. The McKeen carried twenty-three passengers and a crew of two, though one operator could and frequently did suffice. In any case, the "jump car" was far ahead of its time. Its patent pre-dated those of the Westinghouse single-unit coach by five years, and at the time the Frisco's was reputedly one of only three McKeens in active service.

In the local language the unit car was no great shakes. Its "backwards" or reverse engine remained definitely unpredictable; the two engines together could not maintain scheduled speeds on heavy grades or in rainy or snowy weather. According to the local depot pundits, when the McKeen had one engine operating, it could frog along on the level, and even with no engine it could run downhill. But two classes of passengers were inevitable. When both engines stalled, the first-class passengers had to get out and walk; the second-class had to get out and push.

Editor-publisher Leon Smith had one part-time reporter, who just happened to have been my older cousin, Alan. Mr. Smith at the time was preoccupied in the print shop, producing "picking tickets" and pickers-wanted notices for the predominant money crop of the realm. At such climactic times Mr. Smith let his reporter do the reporting.

My cousin Alan, however, vowed—in fact, swore—he would be a ——— if he would rouse at 5:00 A.M. to meet any ——— who chanced to be arriving on the southbound morning train, and the hell with it. In such situations Cousin Alan turned to

me, stoopid li'l Ole Chorley, who was enough of a ——— idiot
to be up and around that early to milk the ——— family cow
every morning before going to school. Consequently, I was
authorized to meet the early morning train, find out whether or
not the renowned visitor arrived, and if he did, arouse my cousin
at a respectable hour, such as, say, 9:30 A.M. I took the assignment gratis, being that kind of a ——— idiot.

At approximately 5:40, the Frisco's Meteor, a mere half-hour
late, came puffing in. From the dismal brick platform I saw the
train come lurching down Wilson's Cut. My grandfather had
contributed the right-of-way only to repent later of his generous
deed when the ——— stinking, noisy trains disturbed his sleep,
his thinking, and his drinking. There were times when Grandpa
Alf stood on the somewhat rickety wooden bridge and slammed
down rocks at passing trains.

One passenger descended from the Pullman coach. He was
elderly, at least he seemed so to me, and he was tall, slender,
ramrod straight, and rather narrow shouldered. He carried a
handsome cylindrically shaped leather bag in one hand, a black
derby and a light raincoat in the other. His dress was elegant, I
thought: his suit was sharply creased blue serge with the double-
breasted jacket, and his black shoes bore a high polish.

My impression of his elegance could have been exaggerated.
I was young then, and Fayetteville, Arkansas, of 1912, was
hickory shirt and overalls country—far more natural and reas-
suring than its current marked-down sport jacket status. But I
remember quite clearly the visitor's appearance. He was exceptionally handsome. His forehead was high, his profile well
shaped. His graying hair was closely cropped, and the vestige of
a mustache was also pale gray. As I remember, his eyes were very
oddly colored, somewhere near that of salted green olives. Like
his well-shaped mouth, they seemed gentle and poetic.

That appearance enabled me, shy and stammery as I was, to
make bold to ask if he were the great engine inventor.

He smiled down at me and nodded. His voice was rather deep.
His words, though quite understandable, seemed tinged with a

slight accent, more French than German—or so it seemed to me, who at that time had known only one immigrant Frenchman, an old-style gardener named Bodeen, and one fresh-over German, Fritz Latouwski, who labored as shipping clerk at the local wholesale company.

I listened closely as the distinguished newcomer stated with thoughtful care, "I am Diesel, but not Doctaire Diesel." He nodded, smiled again, and shook my hand. Then without inquiring into my intentions, he asked about a local hotel.

With considerable stammering I explained there were two hotels, the "regular one" uptown owned by the Fulbrights, who owned pretty nearly everything else profitable in the town, and a little hotel nearby and downtown.

"In walking space, yes?" Diesel asked. When I began leading the way, he skirted the local horse-rig taxi and followed me down the platform in the general direction of Gholson's Inn, the *hoi polloi* hostelry. The visitor, who appeared to limp slightly, turned to scan the station bulletin board which was supposed but frequently failed to list arrival and departure times of the trains.

Failing to find the desired information, Diesel walked rapidly to the passenger window where a Western Union clerk drowsed at a temporarily stilled keyswitch. When questioned, the clerk responded that the McKeen car was due to arrive within the hour and leave around "noontime or a little after . . . ." Diesel thanked his informant and rejoined me for the walk to the hotel.

I took advantage of the interval to explain my unasked reason for having met the train. The engine man smiled again, and his handsome face seemed considerably younger with the smile. He said he would be quite willing to talk with my reporter cousin later in the morning. I suggested that in the event Cousin Alan didn't see fit to rouse out of bed, I should like to interview him. Would he be at the hotel? He answered that he rather expected he would be out in the rail yards with the McKeen car when it arrived.

He paused to view the grim, brick-built box that was the local ice plant—Fulbright's, naturally. He said he would like to look inside, but we found the door still locked. "I am a kind of ice man myself," the visitor explained. "At any rate I used to be, and, in a sense, I still am."

I left him at the dingy little hotel entrance. Time was getting along, and I still had to walk home and do the milking before going to school. I learned en route that Cousin Alan was waking and would "do" the interview. His later unpublished summary was, "The old bastard [Diesel was then fifty-five] asked at least five questions for every one he answered."

When the McKeen car finally arrived, a mere couple of hours late, Diesel was waiting tensely. The self-propelled coach was not too much to look at and a fumy nightmare to ride. But technically it was one of the memorable new facets of American railroading. About a year later, under Diesel tutelage, design engineers at the Sulzer Brothers plant in Winterthur, Switzerland, began work on the first single-unit rail coach powered by a Diesel engine.

Having chatted amiably with members of the local section crew, which was mustering for its day's work, the visitor resumed his erect vigil. When the McKeen car finally came chugging and spluttering to its siding, Diesel waited for the few passengers to descend, then waved to the engineer, and stepped aboard. He quickly made friends with the two-man crew, early vintage automobile mechanics hired by the railroad. He asked many questions, accepting the answers with deference and without comment or debate. Then with great care he began inspecting the twin gas engines. Without regard for his neat blue suit, he knelt on the gravel ballast and then lay flat—"like a damn wallowing horse," my Cousin Alan, who had by now arrived on the scene, inserted—to look over the wheels, chassis, and visible transmission parts. He presently regained his footing, thanked the crew, and then turned to my tall, nineteen-year-old cousin and inquired, "What other interesting things are there to see here?" When my cousin was sarcastically vague with his answer, Diesel asked what the town's principal interest was. My cousin answered, "Waiting for relatives with maybe a little land or money to die off."

On continued interrogation Alan recalled that a merry-go-round was being set up in Trent's cow pasture north of town. The term seemed to baffle the visitor, who presently asked if it could perhaps mean the same as "carousel." Cousin Alan replied that he didn't know what "carousel" meant.

At that point, the station master approached. He had arranged for the benefit of the distinguished visitor a short McKeen car tour of the St. Paul branch of the ever astonishing Frisco Railroad. The tour was quite short, but so was the St. Paul branch. It rambled and twisted through some twenty miles of what still rates as among the most primitive backwoods hill country in all the nation.

St. Paul, Arkansas, was then the white-oak capital of the nation, you might almost say of the world. Its output of crossties underlaid railroads on all continents, and its woodsmen, tie-hacks, were as colorful as any who ever raped virgin hardwood. Appropriately, the tie-hacks were train riders, though usually only open logging flats. In enclosed coaches they were a peril; they sipped moonshine from earthen jugs or glass fruit jars, danced jigs on the car seats, smashed windows, and often sprayed the passing countryside with buckshot or six-shooter slugs.

Although he had no opportunity to indulge in the perilous pastime of train riding with the tie-hacks, Diesel saw the wondrous Arkansawyers in person. He recalled with lingering astonishment their striped silk shirts and their facility with the banjo, an instrument which the inventor had apparently never encountered before.

A very few minutes after school was out at four, Freddy Drake, a school mate, and I arrived at what Diesel called the "carousel." Shortly Rudolf Diesel came strolling across the freshly green countryside. To my surprise Cousin Alan was serving him as guide. The visitor approached the still inactive merry-go-round, and quickly made friends with the fat old man who owned the rig and the lean young man in the frayed suit and shiny hat marked "Conductor." As the two went back to assembling the rig, the inventor examined the circular railing and the midget-flanged wheels on which the revolving platform rode. "Like a steam engine railroad," he said.

"Steam *engine*," squeaked the skinny young man.

"Plant!" Diesel insisted. I could not hear his explanation of the term, but it was rather lengthy. He recognized me on second meeting and chatted pleasantly with Freddy Drake. "There are two sons and a daughter in my family."

I replied that I came from such a family.

"You are of a farmer's family?" he inquired.

I nodded. He said that he had relatives who were farmers, including a second cousin in Texas.

"You help your parents with the work?" he asked.

"Yessir."

"That is good. Boys should help their fathers." He asked if my family's farm were anywhere near. I assured him that it was within a quarter of a mile. Diesel lifted a large gold watch from an inner pocket, viewed the timepiece, then glanced down at me, and explained that since his train would not leave until eight that night he had some free time.

I asked if he would like to visit our farm. He nodded and said he would like to see an American farm. Without further conversation and with Freddy Drake and Cousin Alan tagging along, he walked with me to my home. In the front yard, he met my father and mother. As things turned out, he did not look at the farm.

Before "taking up" farming, my father had been a partner in a wholesale grocery firm. He happened to mention that some of the big traveling circuses had been among his best customers; and that in dealing with some of the "big tops," such as Ringling Brothers, he had met German army officers who were traveling with the circus. The Kaiser's army had recognized the itinerant American circus as a prime example of efficient transporting, feeding, sheltering, and otherwise servicing great numbers of people. The Imperial Staff, therefore, had assigned select officer personnel to travel with American circuses and learn firsthand how they were provisioned and "disciplined." Diesel mentioned that he had once lived in Berlin as a neighbor to various members of the Imperial General Staff. He added somewhat curtly that as he had known them, most of the Kaiser's "war men" would do better in circuses than anywhere else.

Then a neighbor who was a second generation Bavarian came a-borrowing. Adolf Franz almost instantly recognized the caller. "You are Rudolf Diesel! I recognize you from your pictures . . . . You are German."

"Bavarian," the visitor corrected. After a brief exchange in German, Diesel said in English, "I take it you came from Bavaria, too." He spoke to my mother apologetically, "You should forgive us. We are only a pair of funny-paper characters—Adolf and Rudolf."

My mother, the family's irrepressible diary-keeper and facts-checker, missed the point. "You *are* German, aren't you?"

The inventor ceased smiling. "I am one from everywhere," he answered. "My paternal ancestors were Slavs. I am French born and Bavarian naturalized. I like to be a citizen of all the world." He spoke briefly of his parents, both "internationals." He mentioned that he had always been surrounded by intelligent women, including his mother, sister, wife, and daughter. "Only *I* seem to stay stupid."

After more conversation and adjournment to the front porch, Diesel again looked at his watch and suggested that it was time for him to go back to his hotel and get ready for the overnight trip back to St. Louis. He graciously declined an invitation to stay for a "snack of supper," explaining that he was obliged to take medicine before meals and that his medicine kit was at the hotel.

At that point, Herr Franz, who was also one of the first three automobile owners in our neighborhood, volunteered to drive the visitor back to his hotel. Diesel accepted. Herr Franz drove the car, roaring and dust-splattering, out of his garage. Before climbing into the open auto, Diesel told a brief joke in German.

Adolf Franz laughed so hard he almost smashed into the big maple tree. He later declined to translate the joke.

The northbound night train used to be the town's best attraction. That night I was in the crowd. Diesel smiled at me; then somewhat impulsively, he opened his cylinder-shaped bag and handed me a bottle of soda pop. "This is the American drink I like best," he declared. He took a multiple-bladed knife from his jacket pocket and pried off the bottle cap. "This kind they call cream soda is best," he stated with great conviction and handed me the bottle, explaining that he would wait until bedtime to have another for himself.

At approximately that point, another of my numerous cousins arrived. He was John Ellis, an engineering student at the local college. He was also something of a prankster, having to his credit such revered academic achievements as dismantling the horse's skeleton in the veterinary shed and reassembling it on the upper front porch of the girls' dormitory. He had been formally expelled but informally reinstated in the state's sole university after he had tamped the memorial cannon with blasting powder, loaded it with old shoes collected from the boys' dormitory, wadded the charge with his brother's underwear, and fired the charge of footwear across the historic front campus.

On this particular evening, John Ellis was sporting a trick straw hat equipped with a side rigging of rubber tubing which ended in a squeeze bulb. Squeezing the bulb tipped the hat. Under ideal circumstances, it would also squirt water or other fluids on the person saluted. Cousin John squirted me slightly, but did not molest the distinguished visitor. Diesel examined the hat with care, then smiling faintly, he handed it back without comment to the owner.

As usual the train was late. Rudolf Diesel felt like talking. He talked about railroads, especially American railroads, and the interesting things one could learn from train trips. He paused to direct that I go ahead and drink all the bottle of soda pop since he had plenty more. He talked interestingly about railroads in France, Germany, and Russia. Then he was most interested in one particular American railroad, the Pennsylvania, which had invited him to serve as a consultant on power devices.

He ceased speaking when the locomotive's whistle sounded its yard call. He watched intently as the reddish-gold headlight came in view around the Y-curve. As the puffing coal-burner came hissing and rumbling past, grinding to its stop, our famous visitor picked up his bag and slipped on his hat and raincoat. Then he glanced down at me and said, "Good night."

He walked briskly up the platform and disappeared into the lamp-lighted Pullman coach. I never saw him again.

Farewell to America

AFTER HIS RETURN TO ST. LOUIS, Diesel's solitary, exploratory trips were limited. The plan for publicizing his visit included an escort and press coverage. The working press of St. Louis joined in the acclamation.

On April 6, following his return from his brief trip to the Ozarks, the inventor rejoined his wife at the Busch home where a reception for the press had been arranged. When the reporters began arriving, there was concern because Diesel failed to appear. A brief search party, which included J. C. Makin, a fledgling *Post-Dispatch* reporter, presently located him far out in the railroad switching yard in deep conversation with J. L. McCarthy, a roadmaster for the Pennsylvania railroad.

The press gathering moved to the reception rooms of the brewery where refreshments were served and "photography arranged." Next afternoon's *Post-Dispatch* carried a photograph captioned, "Dr. R. Diesel, of Munich, Germany, inspects stationary Diesel engines in Anheuser-Busch Brewery." Almost half of all Diesel engines then at work in the United States were in that thriving establishment. But the companion picture, also expertly posed, indicated that fact was only temporary. Its caption read: "Dr. Diesel of Germany visits the new Busch-Sulzer Diesel Engine Works."

With the ground-breaking for the new engine works scheduled for April 14, the drums of publicity were beating lustily. The

Busch executives had wakened somewhat belatedly to the fact that, however famous he was in Europe, Rudolf Diesel was not well known to the American public, even of the St. Louis area.

Background stories mentioned only ten specific Diesel engines then at work in the United States. All these were stationary engines, and all were limited to such prosaic functions as filling and draining brewery tanks, pumping water, grinding stone or cement, or "sucking coal." The fact that the world total of Diesel engines then at work exceeded 70,000, that 365 ocean-going ships—including 53 underwater or submersion craft—were powered in part or wholly by Diesel engines, and that plans were well advanced for Diesel-powered locomotives and dirigibles were generally impressive. The St. Louis press quoted the inventor regarding the future of "clean smoke-free, rational engine plants placed in the cellars of these wonderfully developed American buildings called skyscrapers." The story went on to mention the fact that already Diesel engines were pumping water, generating electricity, and helping heat buildings in several dozen European cities.

The emphasis of the newspaper stories shifted from Diesel's international importance as an inventor to St. Louis's newest engine factory and to human-interest accounts of the Diesels. Diesel meticulously avoided references to his immediate family or revelations of his private life, but his private opinions shone brilliantly. He repeated his belief in permitting every workman to own and profit from a share in his place of employment. He stated frankly that he deplored "extreme nationalism," and he would name himself only as a citizen of the "civilized world." Indeed, yes, the last named included America. He believed that the United States of America was already the leading power on earth and that American industrial and financial supremacy and leadership would keep growing.

"What do you like most about Americans?" inquired Paul Greer, then of the Omaha *News-Bee* but soon to be complete edition editor of the St. Louis *Post-Dispatch*.

"There are four American virtues which I believe basic," the inventor answered. "First, I would name the absence of any durably fixed classes. Your poor and rich, laboring and professional classes are quite readily interchangeable . . . . Next, I would say that the distribution of intelligence and charity among everyday Americans is very fair and generous. My third point is great respect for American inventive talents, also exceptionally well distributed among all kinds of workers. And finally, the modesty of important Americans. Modesty of important people invites more of its kind."

A question concerning his views on "internationalism" was clearly loaded since the Diesel engines were already in extensive use among great nations which had been and quite possibly would soon again be at war with each other. "I detest all war," the inventor declared. "I am a passionate pacifist. My aim is to build power for peace—not war."

Asked about his "American ambition," Diesel smiled playfully. "To drive my own thermo-locomotive from New York to St. Louis—fueling with nothing but butter—if you will kindly spare me the butter."

Questioned seriously about his immediate plans, Diesel answered that he had scheduled some ten thousand miles of travel on American railroads to "glimpse" at least twenty states. "I like to travel by train. I find your monster American railroads very exciting."

He added that he had scheduled three more public lectures: one at the United States Naval Academy, another at Cornell University, and a final appearance at Cooper Union in New York City. He had, further, accepted an invitation to attend and exhibit his engine at the forthcoming San Francisco world's fair (then scheduled for 1915), and he would be a guest of the superintendent of the United States Naval Academy on an inspection tour of the Panama Fair, which was being planned to mark the completion of the Panama Canal in 1914.

Would he be returning to St. Louis?

Indeed, he would. He had already accepted American employment—as technical adviser for the Busch-Sulzer Diesel engine factory already being built in St. Louis.

A *Star* reporter asked if Diesel had plans for "naturalizing"— becoming an American citizen. The inventor answered that he would regard American citizenship as a very real honor; however, he was obliged to take his future steps one at a time. Important work waited his return to Munich and Europe. It included supervising various phases of building and opening the new Diesel plant in Ipswich, England.

In his final St. Louis interview, Diesel again declined to comment on the rumor that plans were under way to power components of the British fleet with Diesel engines. He reiterated that he was not in direct negotiation with any military or naval group or power and added that Diesel engines were already being used in shipping under at least eight different flags. The ascendancy of Diesel power for shipping was inevitable. Whether on land or sea or in the air (Diesel had long since designed engines to power the Zeppelin airship), the "reasonable function" of the rational heat engine would remain for the "business of peace." Rudolf Diesel was not and would never be a party to any nation's preparations for war.

In their thirteen days in St. Louis, the Diesels were the subjects of many society page headlines. At least three social events were sponsored by the local Lutheran church. Martha Diesel was surprised, and her husband was quietly amused, by the press buildup of Martha's eminence as a leader in Lutheran women's work. While being entertained by the Reverend and Mrs. E. L. Ruppel, Lutheran missionaries on leave, Diesel listened in good grace and with straight face while his wife was introduced to visiting society reporters as an "internationally revered leader in Lutheran women's endeavors." He was seen shaking his head when the next day's issue of the *Post-Dispatch* described him as a devout leader in Lutheran men's endeavors. The latter statement was gratifying but wholly incorrect. Two days later when the Busches entertained with a "high tea" honoring both the Diesels and the Ruppels, Rudolf Diesel was not present. A reporter covering the event for the *Post-Dispatch* wrote a glowing half a column with the caustic closing comment: "Notable by his absence was the senior guest of honor."

By and large the inventor and his wife fared well in St. Louis. They had been graciously and warmly received by an old town young again with springtime. The Diesels left in good spirits. Beyond doubt both were beginning to like America and Americans.

The three major lectures turned out well. The first, an appearance at Cornell on the evening of April 18, was the "revered university address." Wisely, Colonel Meier and his Busch-Sulzer colleagues had accepted the invitation of the engineering faculty and students of one internationally respected university instead of several from a succession of lesser schools.

The inventor arrived at Ithaca with a painful and badly swollen right foot. His gout had "flared." But both the Diesels could take comfort from the fact that they could have been infinitely worse off. On arrival at Ithaca they encountered blaring headlines telling of the *Titanic's* catastrophic collision with an iceberg off Newfoundland. More than three-fourths of all aboard, 1517 by the best estimates, were lost at sea. The odds were overwhelming that had the Diesels been aboard as they had hoped, they would have been among the casualties.

The Diesels had sought well-earned rest in a Pullman drawing room en route to Ithaca. Rudolf may well have anticipated trouble. In any case, he arrived in great pain from a severe seizure of inflammatory gout. Almost overnight his right foot had swollen to half again its normal size. The student reception committee was obliged to search several dormitories to find a black shoe big enough to fit the swollen foot. Haggard and drawn with pain, the inventor donned his white tie and tails and hobbled to the podium where he valiantly delivered his lecture.

After convalescence and brief medical treatment in New York, the Diesels entrained for Annapolis. Diesel's address of April 26 to the corps of midshipmen went quite well and much more

Courtesy M.A.N.

M.A.N.'s first Diesel automotive engine, without air compressor and
with direct fuel injection, developing 45 horsepower
at 1,500 rpm, 1923–24.

Courtesy R. G. LeTourneau,

Diesel-powered earthmover for use in canal building—earth load, 130 tons; engine rating, 18,000 horsepower. Designed and built by R. G. LeTourneau, Inc., Longview, Texas.

Courtesy Caterpillar Tractor Co.

Caterpillar Diesel D8 tractor with Trackson pipe layer.

Courtesy R. G. LeTourneau,

Diesel-electric tree crusher. Designed and built by
R. G. LeTourneau, Inc., Longview, Texas.

Courtesy Caterpillar Tractor Co.

Diesel engine in use at a Grand Coulee Dam
irrigation pumping station.

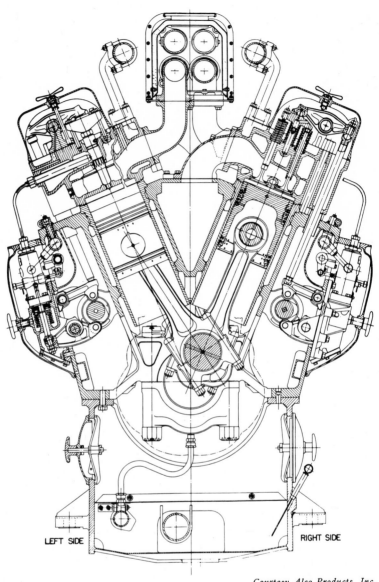

LEFT SIDE   RIGHT SIDE

Courtesy Alco Products, Inc.

Power-placement diagram for a modern Diesel locomotive.
Cross section of V–251 engine, looking from generator end.

Courtesy Nordberg

The largest Diesel engine built in the Western Hemisphere—a two-stroke-cycle, twelve-cylinder Diesel engine of 12,800 horsepower at 200 rpm.

Courtesy M.A.N

Rudolf Diesel, successful inventor.

comfortably than his Cornell lecture. Before another prestige audience, Diesel discussed, appropriately, marine engines and their influence on sea power.

The final lecture, on April 30, before the convention of the American Society of Mechanical Engineers, was perhaps the most brilliant of his American lectures, certainly the best publicized. New York's Cooper Union auditorium was crowded, with only standing room remaining. The two-hour address was followed by a prolonged ovation, a standing vote of appreciation, and an impressive flurry of newspaper stories. Some of the stories noted that the *Titanic* had *not* been Diesel powered. The inventor did not dwell on the point.

Visibly elated and deeply comforted by the exceptional success of his Cooper Union address, Diesel recuperated briefly in New York City. On May 3, he and his wife left for another unique American experience, a meeting with the most famous of American inventors, Thomas Alva Edison.

We are indebted to Diesel's son Eugen for the description of the visit in Orange, New Jersey. He recalls that his mother and father were greeted by the famous inventor, dressed in work clothes, in the modest cottage which also served as his laboratory. The visitors arrived early in the morning, and their host led them to a simple breakfast being served in his workroom.

The entire house centered about the large, cluttered workroom. The walls were lined with wooden partitions. A metal cot stood in a slightly less crowded corner. A roll-top desk was tremendously crowded with papers and drawings. Tables cluttered with models and parts were crowded in the center of the room. Dozens, perhaps hundreds, of models, for the most part fragmentary, were in easy reach. The visitors noted that many of the models were of cement houses. Apparently the inventor was temporarily on a cement spree. The oversized and only visible armchair in the workroom was made of cement. Over all the disorder, austerity prevailed.

Within the first hour, Diesel realized that he and his host were not as compatible as he had anticipated. Edison, then sixty-five, was eleven years older and, as Diesel felt he indicated, several lifetimes wiser.

The first day Edison reviewed the highlights of his recent trip to Europe. His wife had "dragged" him to Italy. Italy had bored Edison profoundly. To begin, and apparently to end with, there are too all-fired many churches in Italy, and Edison simply wasn't interested in churches. His wife kept dragging him to see cathedrals and such like until presently he put his foot down and refused point-blank to go into even one more church. As for murals, high art works, tapestries, sculpture, and all such as that, Edison had at best a faint interest in them. He found some of the so-called holy art absolutely revolting. Take the painting of St. Sebastian, for example. That Sebastian was nothing but a human pincushion!

Edison had what Diesel could only term an encyclopedic mind. He knew a little, but a pertinent little, about practically any subject anyone could mention, including the Diesel engine. The American stated that he was "self-educated." His scholarly visitor could only regard this as a somewhat dubious asset. Diesel, after all, had graduated with exceptional honors from a German technical school with standards far above those of the average American college of the time. More important, he was a scholar with broad interests and training—a mathematician, physicist, thermal engineer, and philosopher.

Edison, whom Diesel had at first regarded as an imaginative electrician, rather waspishly denied confinement to any "speciality." Diesel considered himself inadequately learned in the area of things electrical, including the new electronics. But he regarded Edison as inadequately learned in both the absolutes of inventive competence—mathematics and physics.

Another basic conflict centered in the respective concepts of "practicality." Edison regarded an invention as practical when, if, and because "it works." Diesel's definition of practicality was based on productiveness. As an inventor of engines, he had given the world power-producing machines rather than power-consuming machines such as the electric motor, the incandescent lamp, the motion picture projector, the phonograph, and so on. He respected the utility of many of Edison's inventions, but he also believed that many of the amazing conglomeration were merely playthings.

Edison regaled his visitor with his views on engines—expositions which to Diesel were hardly more than froth on beer. Edison imagined what he termed an "accumulative" engine—to be developed by the systematic "compositing" of many different scientists and engineers. Diesel's failure to comment was his usual manner of showing disapproval. When the host repeated his confidence in the "composite deal," whereby the work of a dozen or more engine makers might be brought together in one "light and practical machine," Diesel broke his stoic silence by asking if Mr. Edison by any chance considered engine makers worth a dime a dozen? Edison seemed to enjoy the thrust. He drawled that his real concern was whether "what a dozen engine inventors bring together [was] going to be worth a dime when brought together."

The visit was to have lasted a week. One can hardly fail to be astonished that it actually did last five days. Credit for the accomplishment must go to the Mesdames Edison and Diesel. "The girls appear to get along just fine," the host reflected.

Diesel was plainly disconcerted by Edison's repeated assertion that a successful inventor is primarily a successful co-ordinator of the works of other inventors. Diesel's contention was that a successful inventor is one who creates challenges for skilled colleagues and competitors by providing them a foundation for worth-while work. Each evening Edison took time to leaf through the most recent reports of his various employees or co-workers and from the progress reports to decide on and assign or recommend the next project. Diesel's practice was to tackle a problem directly, arrive at his own solution or verdict, and hand it along to others to accept, reject, or improve on.

Their one point of open agreement was dedication to their work, to which both were devoted slaves. Both kept indefinite and almost continuous working hours. Both were bed-table workmen.

Otherwise the two famous inventors had astonishingly little in common. Long before the five days had passed, Diesel had Edison securely tagged as one of the better-publicized American curios. Apparently Edison had classified Diesel as a neurotic bohemian who belonged in saloons and art galleries or possibly in opera boxes.

The Edison lectures on the Simple and Natural Life recurred with monotonous repetitiousness. The lectures were reinforced by a daily meal of "fresh vegetables flavored with their own juices." Diesel also was mildly annoyed by Edison's near fanatic disdain of all things alcoholic. To the host's assertion that "swilling alcohol is unnatural," Diesel could only point out that according to the little he knew of history and anthropology, *Homo sapiens* had been consuming alcohol almost as long as he had been consuming bread and meat. When Edison demanded to know why his guest said "almost as long," Diesel smilingly answered, "Only because fermentation takes a little time."

When taking his not regretted leave, Diesel remarked caustically that Edison was truly a glutton for the simple life. As the Diesels left, Thomas A. Edison called out, "Don't eat so much!"

On May 14, after a few days' convalescence at the old Waldorf Hotel, Diesel and his wife traveled by rented limousine to the Hudson River pier only a short way from the western end of Thirty-fourth Street. With haste to avoid photographers, the middle-aged couple boarded the *Berengaria* and sought solace in a first-class stateroom. Shortly after noon, the deck steward and quartermaster began calling all ashore. The ship's band opened its concert in the State Lounge.

Music-loving Rudolf Diesel was not on hand. According to his wife, he limped to the stateroom window and watched the New York skyline fade from view. He liked this America. But he would never see it again.

"En attendant je t'aime"

 THE DIESELS WERE MET on their arrival at their Maria-Theresia-Strasse villa with glad tidings of another arrival. In Frankfurt their daughter Hedy had given birth to their first grandchild. Arnold von Schmidt was understandably and almost deliriously happy.

Rudolf Diesel's happiness was near jubilation but, alas, short-lived. The month's end found him afflicted with financial woes almost as painful as his headaches.

The most immediate cause was the worsening fortunes of his real estate speculations. Heilmann, a prominent Munich real estate firm, proceeded to make good its prior threat to enter civil suit against the inventor. After complex litigations which stretched over two years, Diesel lost the suit. Even before the final deposition he was forced to pay a claim of some 600,000 marks in cash. Before the first stage of the suit was ended, he had been obliged to discharge most of his servants. To pay the settlement, he had no recourse except to mortgage his home. By summer's end his ready-cash reserve was completely exhausted.

As money worries multiplied, the inventor returned to frantic and in some part ludicrous attempts to allay his successive financial failures, which by midyear 1912 totaled nearly ten million marks—very nearly two and one-half million American dollars. Desperately Diesel used his dwindling credit to cover losses. Frantically he bought into and soon became the principal shareholder of a corporate distributorship for electric automobiles—"electromobiles." This investment, too, proved to be a dud, paying no dividends and caused painful losses. Once more he quarreled with former banker friends and spurned others who might have given helpful counsel.

His headaches returned to torment him; his gout remained chronic. His face grew more haggard and more deeply lined. He was suffering long periods of deep depression and moodiness.

The one promise of hope was the year's end publication (1912) of his full-length definitive history of the Diesel engine—*The Origin of the Diesel Engine (Die Entstehung des Dieselmotors)*. This work, which reviewed in careful detail the crucial first four years of construction, had been germinating for several years, but it grew directly from a scholarly lecture which the inventor had delivered to the Shipbuilding Technical Union (*Schiffsbautechnischen Gesellschaft*) in Berlin (November, 1912). The gist of the argument was that an invention consists of two distinct parts, the primary idea and its practical development. The "high ideal of the inventive proposal" is seldom achieved by the final product, but much of value is learned in the long process of creation.

A wearying inventor plodded into another wearying year. Somehow his heavy steps seemed oddly interspersed with lighter ones. Diesel began giving more time to visiting old friends in various cities and countries—Paris, Berlin, Amsterdam, and so on. He was more sympathetic and responsive to friends than he had been during the previous decade. He showed a renewal of interest in art and in travel.

Early in 1913, he and Martha traveled to Sicily, island of eternal spring. From Sicily, Diesel wrote: "We can say good-bye here, we will not see it again." Before returning to Munich, they made brief visits to Capri, Naples, and Rome. While mountain winter lingered, Diesel made still another pilgrimage to the Bavarian Alps, and once more he climbed the Säuling. His guide reported that he remained an unusually long time atop the mountain, which overlooks white-clad fellow mountains. The inventor seemed completely lost in deep contemplation.

He continued the holiday alone, traveled again in the Bavarian Alps which he and his daughter had visited so happily before Hedy's marriage. Alone he viewed again the fairy-tale castle once called Neu-Hohenschwangau, built by Ludwig II. He again saw the Sängerhalle, a copy of the original setting of the saga of Wagner's *Tannhäuser*. Once more he looked down on the Allgäu countryside he loved so well. On a far horizon he could see Swabia, homeland of his ancestors.

Diesel found time for other journeys to visit old friends and familiar places. He visited his sister, Emma, in Ragaz. He went farther to revisit Winterthur and the Sulzers. There Frau Sulzer-Imhoof noted perceptively, "That wasn't the proud Diesel any more."

Sentimental moods were interspersed with brief and unusual flares of temper. Returning to Munich where his younger son Eugen was attending technical school, Diesel managed to present his would-be follower a check for ten thousand marks for deposit in the *Reichsbank* as a "further studies fund." Some days later when Eugen told his father that he had deposited the check with another bank, the *Bayerische Hypotheken und Wechselbank*, Diesel was momentarily furious. Yet the inventor himself had patronized the *Wechselbank* for many years. Perhaps his reaction came from a feeling that the bank he had once believed a "pro-Diesel" had become "anti-Diesel."

While nurturing some prejudices, the inventor was shedding others. For example, he had determinedly shunned travel by any kind of aircraft as too dangerous. Yet at Leipzig, where the *Internationale Baufach-Austellung* was in progress, Diesel willingly made a trip on the dirigible *Sachsen*. Previously he had refused to participate in such flights because of the extreme fire hazard. This time he willingly took the risk and, from all appearances, enjoyed the adventure immensely.

There were other gratifying events. These included a renewal of the offer of employment as a consultant for the Pennsylvania Railroad (which Diesel presently declined) and a rather vague overture from an American automobile maker named Henry

Ford. Again the apparent Diesel reaction was that routine employment, even if validly related to his engine, could not possibly solve his financial problems. He did, however, welcome an offer from the Diesel engine factory at Ipswich (England) to serve as technical consultant for a retainer fee of £1,000, then about $5,000 a year.

By June of 1913, Diesel was again at his Munich home, busily and apparently quite happily serving as a gracious host for a series of receptions for American engineers, including Colonel Meier from St. Louis. In the course of these pleasant gatherings the host confided to one of his favorite guests, George Carels of Ghent, that he, Diesel, felt himself only temporarily in financial straits. "All that really bothers me is being so heavily in debt." The inventor added, "As for being out of or short of money, that's an old story with me."

Carels wrote his brother, Jean, "There is nothing really flighty or inconsistent about Rudolf. . . . I have known him intimately since we were both very young men. He has the calm thoroughness of a high-class German, the culture, modesty, and self-control of a high-class American. . . . He . . . insists he has at least twenty years of work still to complete and is straining at the leash to get on with it."

From Philadelphia, Colonel Meier took time to write Diesel a sincere thank-you letter. "In a long and foot-loose life you are the most gracious host I ever met."

Another grateful guest at what Meier had termed "our wonderful little June shindigs at Diesel's house" was Sir Charles Parsons, at the time one of Britain's most-renowned ship designers. Sir Charles noted with concern that Diesel, his cherished dinner companion, was distressed with financial worries and "inclined to be concerned about the breath of war which he feels to be blowing across Europe." While at Diesel's open house, Parsons made a dinner engagement with Diesel for the following October 1, in London. "I will enjoy your monkeyshines," Diesel told his titled guest, thereby exhibiting another phrase from his newly acquired American vocabulary.

The summer continued to be disturbed by financial worries and heated skirmishes with creditors. Again and again the inventor was painfully and, at intervals, conspicuously upset. His melancholia was increased when his occasional appearances as a lecturer were not received as cordially as before. The spectre of poverty grew less spectral, more real. Records show that as of October 1, 1913, Diesel owed heavily for interest on loans. He had no cash on hand. All of his bank accounts were completely drained. He was deeply in debt.

Rudolf Diesel's intervals of extreme melancholy were lengthening and darkening. It might be ventured that this was in some part hereditary. His grandfather and his father, each in his own manner, had withdrawn from reality. His eldest son, then living in Berlin, was becoming a recluse, even though he was married and about to become a father. To his father's dismay and chagrin, the younger Rudolf Diesel had showed no interest in any kind of machines or physical science or engineering. In a fervent desire to be independent, he had withdrawn from school, and at nineteen left his parents' home to live alone in Munich and earn his own livelihood. His subsequent ventures were far from impressive. He worked successively as an errand boy, store clerk, and later as a clerk for an automobile club in Munich.

The inventor's younger son, Eugen, set out purposefully to follow in his father's footsteps. After basic schooling, Eugen spent a year at Sulzer Brothers at Winterthur, and then returned to Munich to study at his father's alma mater, the *Technische Hochschule.* There Eugen became interested in geology and later economics. Following his father's death, he became a merchant, first in Stockholm and then in New York. During the late 1920's, Eugen took up the study of philosophy and began writing books.

Despite his heritage and his own despondency, there is no responsible testimony available that Rudolf Diesel had ever predicted, threatened, or even mentioned suicide. He had occasionally confided his financial distress. He had repeatedly cried out that in the "mausoleum" of a now heavily mortgaged mansion, he felt like a "living corpse." But he dedicated his final and most definitive book, then in preparation, to his wife, Martha. In the dedication he vowed, ". . . if you were not living, I could not live." The converse was at least implied.

For the most part, August, 1913, was a sultry, rather uneventful month. It found Munich's editorial writers and various pundits speaking more and more darkly of the prospects of a European war.

By September, Diesel's hopes began to rise. Ground-breaking ceremonies for the new Diesel engine plant at Ipswich, England, were planned for early October. Appropriately, Rudolf Diesel had been invited to be present as a special guest of honor. The inventor's retainer from the British Diesel engine company was modest enough, but the honorarium of one thousand pounds yearly, which was to begin October 1, assured that the Diesels would not go hungry.

George Carels, the manufacturer from Ghent, had urgently invited the inventor to visit him in Ghent and from there to go with him to London for the directors' meeting of the British Diesel Company and thence to Ipswich for the ground-breaking ceremonies.

Through many years, Carels had emerged as Diesel's most intuitive friend. The Belgian seemed to sense the immediate need of helping Diesel break away from his long and fatiguing battle with worry. With that in mind, Carels had maneuvered successfully for an invitation from the Royal Automobile Club of London for Diesel to be the principal speaker at the club's meeting on September 30.

Diesel himself was no longer an automobile owner. The previous year he had sold his fiery red NAG with the brass radiator, which, according to gossip, he had driven at reckless speeds on nearby roads and even side streets of Munich. If rumor was true, as it may have been, such behavior was conspicuously out of character since Diesel was usually a careful and considerate driver. In any case, the matter was now of no real consequence.

Diesel had sold his once magnificent auto because, as he petulantly stated, "of the nasty [debt] load which rests on us."

That, too, just was not like Rudolf Diesel.

Early in September when Martha set forth to Remscheid to visit her mother, she planned to meet her husband in Frankfurt, at their daughter Hedy's home, and there join him for his trip to England. Apparently these plans were changed; when the inventor started for London, it was as part of an all-male threesome, George Carels, his engineer Alfred Laukman, and Diesel.

Martha was little more than out of sight when Diesel gave the few remaining servants a long week end and put his youngest son on the train for the Sulzer Brothers plant in Switzerland. After he was alone in his handsome house, he asked his elder son to spend a few days with him. Thus the company consisted of a temporarily depressed father and a habitually depressed son. Diesel brought out the collection of household keys and painstakingly explained what each opened. He directed young Rudolf Diesel to try to open the doors, drawers, and lockers which the various keys fitted. The inventor then pointed out where all his valuable papers were kept.

On an outing to the Starnberger See the conversation turned to suicide. Quite possibly the topic arose from the fact that in this lake Ludwig II had drowned himself. Young Rudolf offered the opinion that the easiest way to end one's life would be by jumping off a fast-traveling ship.

After his son returned home, the inventor remained alone in his great villa. When the members of the household staff returned from their unexpected holiday, they noticed that a considerable quantity of the inventor's papers had been burned in the downstairs furnace. Such an act seemed odd. There had been no chilly weather or other apparent reason for building a fire. The master of the house seemed depressed, but his words and actions were gracious and kindly, even if somewhat evasive.

Presently he packed his bag and boarded a train for Frankfurt. There he spent two weeks, apparently quite happily, visiting his daughter and son-in-law, admiring his newest grandchild and playing with his first granddaughter, by then a charming toddler.

The fond grandfather also visited the renowned Adler *Werke,* where his son-in-law, Arnold von Schmidt, was one of the new directors. Adler built automobiles and stationary Diesel engines and was effectively experimenting with constructing Diesel engines to power automobiles.

In strongly pro-Diesel company the inventor took the driver's seat of an Adler test car and proceeded to drive it at what Martha, who by then had joined the family group, described as a most reckless speed. And that simply was not like her Rudolf.

Otherwise, the inventor seemed in an unusually gentle and affectionate mood. He extended his visit until September 26, taking leave of his young grandchildren and his daughter with great tenderness and paying all due respects to his son-in-law and his parents.

His parting with his wife was most affectionate. In Frankfurt, Rudolf had bought a present for Martha, a very elegant overnight traveling case. He pleaded with her to take good care of the gift and not to open it until the following week. Martha Diesel had become accustomed to observing her husband's special requests regarding his sometimes surprising gifts.

Diesel chose to travel from Frankfurt to Ghent by slow train, but first class, the sole occupant of a luxurious compartment. The rail route followed the legend-rich Rhine River. For reading matter he took along a large volume of Germany's philosopher of pessimism, Schopenhauer, *Parerga und Paralipomena,* a Latin title with a German tinge—*An Extra Ornament and Things Omitted, Not Related.* The solitary traveler left a bookmark between pages which dealt with the handling of wealth, capital, earnings, and estates. Diesel may or may not have recalled that the same Arthur Schopenhauer had written, in *Studies in Pessimism,* that "a certain amount of care and pain or trouble is necessary for every man at all times. A ship without ballast is unstable and will not go straight."

Diesel arrived at Ghent late and went directly to the elegant

Hotel de la Poste, located next to the opera house on the Place d'Armes. There he wrote a letter to his wife. It was a strange, confused, poetic, and beautiful letter, suggesting at least by indirection that its writer was ill. There were lucid sentences, however. Diesel mentioned that he had telegraphed ahead to reserve a room at his favorite London hotel, de Keyser's, and that he hoped for word from her while he was there. Some straying sentences followed, for the most part in German, but with some inter-sprinklings of French. The letter closed: *"En attendant je t'aime, je t'aime, je t'aime. Dein Mann."*

The letter was postmarked between 6:00 and 7:00 P.M., September 27. Diesel addressed or, better say, misaddressed it to her, "p.a. [care of] Freiherr Arnold von Schmidt, Maria Theresiastr. 32, Frankfurt/M."

This was the actual address written on the envelope. The abbreviated street address, of course, was that of his own Munich residence, not of his daughter's home in Frankfurt. Had the letter been correctly addressed, it should have reached Martha the following day. It seems probable that had Martha received the letter promptly, she would have realized that her husband was seriously ill or seriously confused or both. She would have had time to reach him by telegraph or telephone before he left Ghent for his ferry trip across the English Channel. Possibly she would have had time to join him.

But the letter was not correctly addressed. It went first to Frankfurt am Main, where there was no such street address. Postal authorities forwarded it to the other Frankfurt (on the Oder). From there the letter was returned to Diesel's hotel in Ghent. The hotel forwarded it to Diesel at de Keyser's Royal Hotel, London, which reforwarded it to the correct address in Munich, where the letter was delivered on October 5 or 6. Either date was too late.

Also from Hotel de la Poste in Ghent, Diesel had written other letters and postcards to his wife and children. These were correctly addressed, but their texts taken together were quite confusing. The one to Martha spoke of his love for her and told about the activities of the day. The letter to young Rudolf complained of severe headaches, sleeplessness, and bothersome heart.

In any case, Diesel proceeded on his scheduled way. During the afternoon of the twenty-ninth of September, he, George Carels, and Carels' chief construction engineer, Alfred Laukman, boarded the channel ship *Dresden* of the Great Eastern Railway at Antwerp for the overnight Channel crossing to Harwich. Almost exactly forty-three years before, Rudolf Diesel, with a name tag around his neck, had made the crossing in the opposite direction—from Harwich to Rotterdam.

George Carels recorded that the three friends boarded the Channel ferry in good spirits. En route to Antwerp he had sought to reassure the inventor that the Diesel engine surely would soon command a following so strong that the Diesel name alone would be worth many times what he then owed. In view of this absolute certainty, why worry about owing a few million marks?

Carels was completely sincere in his effort to raise his friend's spirits. As one of the better experienced and more successful builders and sellers of Diesel engines, he believed what he was saying. The inventor listened amiably, but without comment.

After boarding the *Dresden,* each man was assigned a state-room to himself. Each saw his luggage delivered, refreshed himself, and then joined the others in the dining salon, where they chose a window-side table and ordered an excellent supper. They dined deliberately and with pleasant conversation. Diesel, according to Carels, ate lightly as usual and barely tasted his wine, but he seemed to be in good spirits. He made no mention of either his illness of the previous day or of incidental aches or pains.

Following the meal the three walked around the promenade deck. They paused at the rail, trying to make out the Schelde estuary and the shore line. The Schelde River was the subject of the painting by Courtens which had deeply appealed to Diesel. Then, by common consent, the friends agreed that bedtime was come. Around ten, shortly after passing Vlissingen, they retired to their staterooms. Earlier they had agreed to breakfast together the following morning. Diesel left a call for six-fifteen.

The night was dark, but there was no rain and the crossing was smoother than usual. All aboard seemed tranquil and routine. But when breakfast time arrived, Diesel did not appear.

Carels went directly to his friend's stateroom. The inventor was not there and could not be found. His bunk had not been slept in, although someone, presumably the room steward, had laid out his nightshirt. Otherwise the Diesel luggage seemed untouched. The inventor's medicine case and glasses were nowhere to be seen.

When Carels hurried to the bridge to report the disappearance to the shipmaster, he learned that a petty officer had earlier found Diesel's hat and overcoat neatly folded beneath a section of the afterdeck railing.

Captain H. Hubert, the ferry master, ordered his craft held at sea for search. A thorough search yielded no trace of the missing inventor. The two-man night watch which had reported "nothing out of the ordinary" had not mentioned the discovery of the missing passenger's coat and hat beneath the afterdeck railing, an incident which must not have been "ordinary."

Still more contradictory of "nothing out of the ordinary" was the disappearance of a passenger. The crew manifest and the passenger list indicated that all others aboard were accounted for; there was no record or testimony of the presence aboard ship of any peace officer, government representative, or anyone else not on the roster. The *Dresden*'s master took pier space approximately one hour behind schedule. He delivered to the port master a formal missing-at-sea certification for the person of the "esteemed inventor, Dr. R. Diesel." He refused steadfastly to issue a death certificate or to comment on the appearance of suicide on Diesel's part.

The German vice-consul in Harwich was notified October 1. Carels and Laukman gave in detail their virtually identical versions of what had happened. Subsequent searching of the ferry yielded no evidence of importance. Diesel's key ring hung on the lock of his suitcase and his steel-cased pocket watch (he had left his gold one at home) was placed so that it could easily be seen from the bed. His notebook was on the table. After the date, September 29, the entry was a small cross drawn in pencil.

Beyond anyone's denial, Rudolf Diesel was missing at sea.

After ten days, on October 10, 1913—the *Coertsen,* a small Belgian chartered pilot steamer, reported having sighted a body floating on the then quite stormy North Sea. According to the records of the port of Vlissingen where the *Coertsen* called on that date, crewmen of the pilot steamer fished out the corpse and took from it some identifying items which were delivered to the port master at Vlissingen.

In keeping with prevailing custom of "finds at sea," the crewmen returned the corpse to the stormy waters. The items for identification included a coin purse, a medicine container, and a spectacles case. Eugen Diesel identified all three of the articles as having belonged to his father. Rudolf Diesel's body was never recovered.

Belgian consular officials and the German consul at London issued a death certificate. The records are not and never were adequate. There was no official investigation or trial by admiralty, not even a ship's company hearing. Since there was no body and no proved status of national waters, there was no official coroner's report.

The overnight bag which Diesel had given his wife before leaving Frankfurt contained German currency totaling twenty thousand marks. Presumably that amount represented all the inventor's remaining cash resources. The money in Diesel's bank accounts had been withdrawn one after another on his own signature. According to Alfred Laukman, Carels had advanced money to cover Diesel's journey to Ghent and to London. The inventor's only readily negotiable property consisted of personal belongings and perhaps some remnants of real estate. The latter, however, cannot be confirmed. Any real property assets were negated by unpaid debts.

The unsoundness of the inventor's real estate investments and

246

commitments had already been aired in civil court proceedings at Munich and elsewhere in Germany. It is ironic and, perhaps, academic to note that, deplorable as some of them were, all or any of Diesel's real estate investments became more valuable than negotiable deposits then on books or in vaults of any German bank through the impending World War I.

Diesel's immediate family took for granted that he had taken his own life. The widow and the younger son openly accepted this theory on the basis of the inventor's ill health, financial distress, and hereditary or otherwise already established behavior patterns. They interpreted his apparently deliberate destruction of private papers in their Munich home and such untypical actions as his airship excursions, his meditative and solitary journeys to old familiar places, and his extremely tender and affectionate good-byes to family members and old friends as preludes to self-destruction. They found additional support for their belief in his deep and persisting melancholia and his long and medically confirmed history of violent headaches and gout. Then, too, there was his apparent inability to stand against ruinous poverty and a proud man's loathing of bankruptcy which could only bring serious injury to others.

The virtually identical conclusions of Martha Diesel and her two sons, Rudolf and Eugen, are admirable for their directness and logic, but they do not erase all the doubt which existed and, to some measure, still exists.

The inventor left no suicide note. He had never mentioned either by word or letter, so far as available records show, any intention to kill himself. Certainly one cannot support the flat statement that Rudolf Diesel had never contemplated self-destruction. The study of psychology indicates that most people do, at times, contemplate suicide.[1]

[1] On the subject of Diesel's alleged suicide, the two authors of this biography differ slightly. From his research in Germany, Nitske is convinced that the death was a suicide and that had there been sufficient evidence to the contrary, Diesel's family and friends would have worked to remove the stigma from Diesel's name.

Wilson accepts the theory of suicide as highly probable but not absolute fact. He admits a reporter's partly intuitive prejudice against freely accepting any

A variety of facts, on the other hand, remain to baffle the researcher:

A family man deeply imbued with the Swabian devotion to continuity of work, Diesel knew that only he—he alone of the Diesel lineage—could carry on the work with the invention which already had changed the fundamentals of man-made and man-used power. By 1913, Rudolf Diesel was well aware that his elder son certainly would not carry on with the Diesel engine and that his younger son most probably would not.

Other facts do not dovetail with the hypothesis of self-destruction. For one thing, Rudolf Diesel, though a prolific and fastidious keeper of records and a man whose whole career was involved with legal procedures, left no will.

Prior to his death the inventor had shown keen interest in taking part in various events of marked importance to his career. He had made—apparently in good faith and in sound mind—a number of specific commitments. These included promises to serve as a consultant for the British Diesel Company and for at least four other companies manufacturing or dealing in Diesel engines. He had accepted many professional and official invitations to attend important meetings in the years directly ahead. The Diesel record for honoring commitments was good.

There is no evidence that Diesel's enthusiasm for his engine or his desire to promote it had waned. His confidence that it would be universally accepted and widely used for an increasing variety of tasks remained strong.

The inventor disappeared en route to a board meeting of a company which he highly respected and on whose board he felt honored to serve. There is evidence that his desire to participate in the ground-breaking ceremony at Ipswich was sincere. Moreover, during his final day, he had confirmed a dinner engagement in London with Sir Charles Parsons, whose company he particularly enjoyed.

---

suicide story. He grants that were odds to be offered, they would be at very least one hundred to one that Rudolf Diesel took his own life. But there is still the one tenuous chance that he did not.

The inventor's American commitments—including the consultancy to the Busch-Sulzer American Diesel Company and its conclave of cherished friends, his promised attendance at the future San Francisco world's fair (Diesel really loved expositions), and his eagerly anticipated tour of the Panama Canal project—had all been undertaken with enthusiasm.

As was to be expected, a crop of rumors sprang up in connection with Diesel's mysterious death, some of them featured in the London newspapers. These, in principal part, were too conjectural or too lurid to justify credence. The various published conjectures that Diesel may have been murdered by paid assassins in the hire of "oil trusts," or rival engine builders, or agents of foreign powers desirous of obstructing his sale of important submarines or naval plans or accouterments, made spectacular headlines. However, no reliable evidence had ever been uncovered by any public or private agency to give any substance to rumor. Rudolf Diesel, who loved people and the peopled lands, died at sea of cause or causes not surely known.

Diesel Engines
Throughout the World

⚍

⚍

⚍ ON THE CALIFORNIA DESERT about fifty miles north of
Barstow and within sight of Death Valley, the National
Aeronautics and Space Agency maintains an outer-space
tracking station. The place is called the Goldstone Echo Site.
Its principal structure is a six-thousand-square-foot parabolic
antenna. The antenna, its space-reaching transmitter, and its
immensely complex console are powered wholly by Diesel engines. They comprise one example of outer space exploration's
dependence on Diesel's rational heat prime mover.

The Goldstone site is far removed from factories or repair
shops. Either the Diesel power works or the tracking station
does not. The same is true of the other moon-tracking stations
at Woomera, Australia, and Johannesburg, South Africa, which
join in providing a 360-degree coverage of the earth's "space
net." Additional Diesel engine power at the Goldstone site
includes eleven mighty Caterpillar Diesel engines, including the
D–353, D–342, and D–320 models. Four Caterpillar D–353's
serve at Woomera and four at Johannesburg.

The United States has a reported 150 underground launching
silos readied for hoisting the Titan II and other intercontinental
ballistic missiles. In each silo, some fifty feet below the earth's
surface, Diesel engines wait to provide stand-by power for
launching the oversize missiles with ranges up to 6,300 statute
miles and speeds of about 15,000 miles an hour. All the engines
chosen and installed by the Strategic Deterrence Force are listed

as White-Superiors 40SX–6's. Their 350-kilowatt generators are General Electric. The Diesel engines themselves are six-cylinder, four-cycle, and turbo-charged with bores of 8.5 inches, strokes of 10.5, and outputs of 510 horsepower (at 900 rpm) per unit.

The role of Diesel engines in the Space Age symbolizes a future without boundaries. The role also provides a vantage point for looking back on the impressive American chronicle of Diesel power.

The first American manufacture of engines which were functionally similar to Diesel's began in 1893. In that year the De La Vergne Refrigerating Company of New York adapted a work model of a low-compression oil engine invented three years earlier by an Englishman, Herbert A. Stuart. Since the engine did not achieve complete ignition by compression, it cannot be properly listed as a Diesel engine. However, after some fifteen years of building engines to power petroleum pipelines, the De La Vergne Company developed a fuel-injection system and so joined the fold of Diesel engine builders. In 1930 the Baldwin Locomotive Works absorbed De La Vergne and directed its distinguished pioneering work to the manufacture of Diesel engines for locomotives, as well as for stationary uses.

The German-American inventor and engineer Carl W. Weiss pioneered the building of low-pressure oil engines in the United States. In 1910 an American sales engineer, Norman McCarty of New York, designed what was then the largest Diesel engine ever put to work in the United States. This was a three-cylinder, four-stroke-cycle engine with a capacity of 525 horsepower at 180 rpm. Its specific use was for generating electricity.

The first "true Diesels" used in the United States were foreign, principally German. One reason for importing engines was the comparatively high cost of producing the engines more or less manually by relatively highly paid American workmen. Many American companies expressed interest in acquiring franchises, but of the first fifty-four overtures made the American Diesel Company, none resulted in the actual issuance of a manufacturing franchise. The manufacturing cost to the American Diesel

Engine Company of a first 50-horsepower unit was $7,955. It was sold to the Anheuser-Busch Brewery for $3,223. In 1898, in order to prove its product, the American Diesel Company had paid the Nürnberg Company $3,300 for a 20-horsepower Diesel engine and sold it to the Morgan Construction Company for $1,800.

Thus at a loss the first working Diesel engines began finding their way into American shops and factories. On September 13, 1899, Seib Brothers Woodworking Factory of Jersey City acquired an engine for $2,400. Shortly thereafter V. B. Cowles and Company of Cleveland acquired one also for regular factory work. A third went to the South Pacific Company of Sag Harbor, Long Island, as a stand-by power source. None of these three performed successfully. The sale of faulty engines was made more expensive by the still greater losses in servicing.

In 1899 the first study of the possibilities for a Diesel-powered American submarine was made. John P. Holland had launched his fifth submarine, *Plunger,* from Baltimore. The 165-ton craft submerged to a depth of seventy-five feet. Under water its two 600-horsepower steam engines generated far too much heat for a crew to endure and survive. The comparatively cool-running Diesel engine seemed promising. Colonel Meier came forward with a small Diesel engine which United States Navy engineers rejected as being too heavy for undersea service. The American Diesel Engine Company then presented plans for a high-speed, light-weight, 500-horsepower engine which might better serve underwater craft.

The plans were sent directly to Rudolf Diesel for appraisal. The inventor called in Hugo Güldner, then his chief engineer, and his good friend from Paris, Frédéric Dyckhoff. After painstaking study, the three agreed that the proposed power unit involved too many hazards.

Late in 1900 a troubled Adolphus Busch called a meeting of directors of the Diesel Motor Company in St. Louis. He advised the group that over $100,000 had been spent by then in the effort to develop and market Diesel engines in America without tangible profit. He announced his decision to close the New York

office. The board decided to sell as much as possible of the unsub-
scribed stock, then totaling $286,000, and use the money to build
an American factory. In the British Isles and Europe, Diesel
engine manufacturing enterprises were already spending many
times that amount for shops or factories. In England, for
example, the newly formed British Diesel Company, headed by
Lord Rothschild, was investing the equivalent of $2,500,000 for
building the new "compressor-combustors."

The American Diesel company reduced its payroll to two
men, but retained its office in New York, while seeking an American manufacturer for Diesel engines. After more than a year one
appeared. The International Power Company of New Jersey,
which owned and operated steam engine and locomotive factories, agreed to invest a quarter of a million dollars in the
formation of the American Diesel Engine Company. After a long
and trying controversy regarding license fees, the new company
put to the test its first American-made Diesel engine, a three-
cylinder, 75-horsepower model in April, 1902. By midyear 1903,
the United States had some twenty-seven Diesel engines in actual
work and sixty-six more under construction. Within another
three years another American manufacturer began building
Diesel engines. The second entry, the Power and Mining Machine Company of Cudahy, Wisconsin, was later absorbed by the
Worthington Pump and Machinery Corporation.

But American manufacture of Diesel engines remained a
profit-seeking enterprise without profit. In 1908, Adolphus Busch
acquired all assets of the pioneer American company and placed
the corporation under the name "Adolphus Busch, Purchaser of
the American Diesel Engine Company." The brewer confided to
his old German friend Heinrich Buz: "We sell many more engines now, but at such prices that when we have installed them
there is nothing left but a bit of glory and advertising." Busch
permitted his Diesel engine company to fade away. Then, on
noting the impressive success of several of the European manufacturers, he made a thoughtful reversal. Early in 1911 he maneuvered the founding of another company in collaboration with

Sulzer Brothers of Winterthur. Thus, the Busch–Sulzer Bros. Diesel Engine Company came into being and a new plant arose in St. Louis.

The new factory shifted quickly to building engines which were patterned after the already proved Sulzer Brothers "K" series. These were four-stroke-cycle, single-acting engines of trunk-piston construction with individually cast bed plates and pressure lubrication of all principal bearings. The four-cylinder engines were built in four different cylinder sizes, bores ranging from 10.5 to 19 inches in diameter and power outputs ranging from 120 to 520 horsepower. They were good, reliable engines, and their manufacture continued until about 1920.

The year 1910 was one of especially notable accomplishment. While the McCarty engine was settling to producing electricity in Casper, Wyoming, the Otto Engine Company, under license from the German Deutz firm, began building Diesel engines at its plant in Philadelphia. During the same year, the New London (Connecticut) Ship and Engine Company organized a subsidiary to build Diesel engines for powering submarines for the United States Navy. Four years later the New London company had produced and installed on the *Maumee* a 2,400-horsepower plant, which was the largest Diesel power installation anywhere. Two years earlier the same company had developed a 300-horsepower unit for powering an oil barge.

About the same time, the Fulton Iron Works of St. Louis began building Diesel engines of Franco Tosi's design. With enlightened determination, the Fulton firm had modernized and expanded its production to include big power plants.

By 1912, Worthington Pump and Machinery Corporation was in the running with what it called the "Snow Oil," a single-cylinder, horizontal, four-stroke-cycle air-injection engine. Six years later Worthington began developing a marine Diesel engine and shortly thereafter a powerful six-cylinder, four-stroke-cycle stationary engine with a rating of twenty-four hundred horsepower. In 1921, Worthington built its first four-stroke cycle, air-injection engine. The progressive climax was reached in 1945 with the completion of Worthington's first supercharged dual-fuel Diesel engine.

Also in 1912, Allis-Chalmers of Milwaukee began building Diesel engines—horizontal, four-stroke-cycle with air injection in units up to six cylinders. About one hundred engines of that type were installed to operate the Oklahoma-to-Illinois pipeline of the Shell Pipe Line Company.

In 1912, too, Fairbanks, Morse and Company of Beloit, Wisconsin, began to manufacture Diesel engines. Two years later, Fairbanks, Morse introduced a series of vertical Diesel engines, built in two-cylinder diameters (twelve and fourteen inches) and in units of one, two, three, four, and six cylinders. This, too, was another memorable first in the American chapter in the Diesel engine story.

The eventual outcome was an impressive variety of Diesel power units. The Nordberg Manufacturing Company of Milwaukee, which began the manufacture of Diesel engines in 1912, continues building a nearly maximum range of the engines, with ratings of from ten horsepower to well over ten thousand.

Other notable American ventures in building Diesel engines began in 1913. One was the Winton Engine Company of Cleveland. Alexander Winton, like Henry Ford, had been a successful bicycle manufacturer who changed over to building automobiles; Winton sold his first motor car in 1898. By 1915 the firm had developed a twelve-cylinder Diesel engine and, during the following year, a six-cylinder, 225-horsepower marine engine. By 1919, Winton had built and installed in his own yacht a Diesel-electric propulsion unit. This installation was a forerunner of a 16,000-horsepower Diesel-electric power plant for the U.S.S. *Perry*. During 1930, General Motors purchased the Winton Company and expanded it as its Cleveland Diesel Division.

Installation of an eight-cylinder Diesel engine in the Burlington Railroad's streamlined "Zephyr" passenger train in 1934 introduced the American era of Diesel-powered railroading. General Motors' Diesel Division ably followed with many other two-stroke-cycle Diesel engines of welded steel construction and unit fuel injection, developing one hundred horsepower per cylinder, in units of six, eight, twelve, and sixteen cylinders. These engines continue to power passenger ships, freight vessels, tankers, harbor and ocean-going tugs, ore carriers, fishing boats, fire boats, dredges, and ferries.

The McIntosh and Seymour Corporation, which was presently taken over by the American Locomotive Company, dates back to 1913. Its first product was a four-cylinder, four-stroke-cycle engine developing five hundred horsepower. Soon the A-frame design was changed to a box frame with pressure lubrication and mechanical injection of fuel.

In 1914, California's Atlas Imperial Diesel Engine Company, which began as Standard Fuel Engine Company, developed a box-frame engine with directly connected air compressors, for installation in a ferry boat. In 1919 the company introduced a solid-injection Diesel engine in which a mechanically operated needle valve was used. This was the first American "common rail" type.

During World War I, electrical engineer Elmer A. Sperry of gyroscope fame became interested in the Diesel engine and developed a mechanical injection system for it.

The Busch-Sulzer company continued its pioneering advances. During 1915, the Navy Department awarded the company the first contract for non-reversible Diesel engines for use in submarines. During the following two years, Busch-Sulzer built about seventy of these air-injection units. After World War I, Busch-Sulzer resumed building earlier designs and added a new two-stroke-cycle engine. This was a single acting, air-injection engine with pistons water-cooled through sliding tubes. The scavenging pump and three-stage air compressor were built integrally with the engine and driven directly from the crankshaft.

By 1920, Busch-Sulzer was back supplying Diesel plants for navy submarines. By the middle 1920's, the firm was building king-size engines—six-cylinder, two-cycle, three thousand horsepower (at 90 rpm)—for the United States Shipping Board. This oversize engine was built for direct connection with the ship's propeller shaft for single-screw drive. It soon became a standard model for commercial shipping.

With the expiration of the company's franchise to build and sell Diesel engines of Sulzer Brothers patents throughout the United States and its possessions in 1926, Busch-Sulzer struck into wide-open competition with a "new" Diesel engine of trunk-piston design, higher speed and lower cost, and of special utility for powering pipelines, ships, dredges, and electric light plants.

In 1930, Busch-Sulzer came up with an all-time standard model. This was a 300-horsepower-per-cylinder plant, with as many as ten cylinders. In 1932 the St. Louis pioneers began producing Diesel railroad engines. The model was V-type, two-stroke-cycle available in from eight to sixteen cylinders, each rated at two hundred horsepower (at 550 rpm). Built largely of aluminum, the engines weighed about twenty-five pounds for each developed horsepower.

With the approach of World War II, Busch-Sulzer produced a total of 304 Diesel engines for use on mine sweepers, fleet tugboats, net tenders, and the like. In 1943, Busch-Sulzer built an additional 112 engines of completely new design for use on army tugboats and other light craft. Later, installation of the Büchi method of engine turbo-charging raised the efficiency ratings of the engines by an astonishing 50 per cent.

In 1946 the Busch-Sulzer Diesel Engine Company was purchased by the Nordberg Manufacturing Company of Milwaukee. By that date, Busch-Sulzer had built a total of 1,670 Diesel engines, ranging in size from 75 to 3,750 horsepower.

The American fortunes and progress of Diesel engines have never followed a single or consecutive trail. Again and again important contributors have entered the competition from various side roads. For example, in 1909, the International Harvester Company began using a single-cylinder, kerosene-burning engine to drive its farm tractors. In 1916, the company began a program of Diesel engine research and three years later produced a single-cylinder horizontal Diesel engine with a pre-combustion chamber and a rating of ten horsepower. Nine years later International Harvester installed its first four-cylinder Diesel engine in a tractor. This pioneer Diesel tractor engine, with a 4.5-inch bore and 6-inch stroke, was succeeded by larger units of the same general type.

Beginning in 1919, the Cummins Engine Company successfully manufactured small, single-cylinder, horizontal Diesel engines based on the patents of the Brons of the Netherlands. The first outlet was Sears, Roebuck and Company, for which some three thousand units were built prior to 1922. C. L. Cummins next developed a simplified fuel-injection system which made possible in America the use of the Diesel engine for powering automobiles. In 1930, a sedan powered by a Cummins-built Diesel engine was exhibited at the New York Auto Show. The following year the engine was installed in a Duesenberg roadster. On the Daytona Beach testing tracks, the test car demonstrated a speed of one hundred miles an hour. The following year the same car entered the Indianapolis Memorial Day Race and completed the 500-mile course without refueling and at an average speed of eighty-six miles an hour.

Test runs continued. An automobile powered by the Diesel engine completed an economy run from New York to Los Angeles; the total fuel cost was $11.22. A Diesel-powered truck run of 14,600 "uninterrupted miles" was next. After having powered a bus across the United States, a six-cylinder Diesel engine was installed in an Auburn automobile and raced at a speed of 137 miles an hour at Daytona Beach.

In 1926, under license from M.A.N. of Germany, the American Buda Company began building an automotive Diesel engine of six cylinders, weighing from thirty-five to forty-five pounds for each developed horsepower. By 1933 the company had introduced a complete line of automotive Diesels.

Meanwhile, the Atlas Imperial Diesel Engine Company, developed a constant pressure fuel-injection system for its four-cylinder Diesel engine which developed sixty horsepower at 600 rpm. In 1928 the Caterpillar Tractor Company installed the

Atlas engine in a tractor. During 1931 the Caterpillar Company built 157 Diesel-powered tractors for farm work and earth-moving. In 1932, Caterpillar hitched one of its new Diesel engine tractors to a twelve-bottom gang plow for a demonstration on a ranch in Oregon. When the tractor stopped after a continuous run of forty-six days, it had traveled 3,500 miles and plowed 6,880 acres at an average fuel cost of less than six cents an acre.

In 1930 the Continental Motors Corporation, a gasoline-engine manufacturer, began shifting to Diesel engines. By 1933 the company was building two types of two-stroke-cycle Diesel engines for the United States Navy. One unit was six-cylinder, the other ten. By 1936 the company had produced a Diesel engine with a rated speed of 2,000 rpm.

The Mack Truck Company became interested in powering their product with Diesel engines in 1927. One of the first tests consisted of installing a Mercedes-Benz truck engine in one of their truck models. A Cummins Diesel was installed in another test model. By 1935, Mack was producing and selling Diesel-powered trucks. It continues as one of the major American users.

More recently, the Le Tourneau Company of Longview, Texas, veteran specialists in oversize construction and earth-moving machinery, swung to Diesel engines. One of the more spectacular results is the gigantic "snow freighter," now widely used in upper Canada and Alaska. This is a huge, oversize loco-motive powered by two 400-horsepower Diesel engines to pull five flatcars each carrying up to twenty-five tons of cargo. The rig carries two thousand gallons of fuel. The power car is insulated and has living accommodations for four men.

Experimentation with locomotives powered by Diesel engines had been carried on in Europe prior to the inventor's death. During 1923 the Ingersoll-Rand Company, General Electric Company, and the American Locomotive Company co-operated in building the first American Diesel-electric locomotive. Out-wardly, it resembled a railway coach. The Diesel engine first used was a vertical, six-cylinder, four-stroke-cycle, single-acting engine of three hundred horsepower. It was equipped with direct fuel injection, accomplished by means of two spray nozzles in each combustion chamber to which oil was delivered under pressure by an injection pump driven from the main shaft. One fuel pump was used to serve all cylinders. In the first test model, the thermal efficiency was six times that of the ordinary steam locomotive—about 30 per cent compared with 5 per cent.

After two years of testing in New York and New Jersey switch-yards, the first Diesel locomotives in passenger train service were put in operation by the Burlington Railroad in competition with the Union Pacific. The first run of the Burlington's "Zephyr" from Chicago to Denver was made at an average speed of 77.61 miles an hour. The locomotive was still in service after twenty-five years and more than three million miles.

In 1941 the Santa Fe and Southern Railways first placed Diesel-electric locomotives in regular freight service. From that time extensive "Dieselization" of American railroads began. By 1955 about 80 per cent of all freight trains, 85 per cent of all passenger trains, and more than 90 per cent of all switching locomotives were powered by Diesel engines. The present figures crowd closer and closer to 100 per cent.

In 1830 (the day was September 15), George Stephenson's steam locomotive, *The Rocket,* traveled between Liverpool and Manchester, England, with a thirteen-ton cargo, at the spectacular speed of fifteen miles an hour. At present, Diesel-electric locomotives regularly draw at least one hundred times that load at speeds five times as great. Nowadays, the Diesel locomotive is recognized by experts as the savior of American railroads, so far as they can be saved.

American railroads, representing one-third of all operative trackage now on earth, have led in the use of Diesel power. The longest American line, the Atchison, Topeka and Santa Fe (13,073 miles of main-line track) has completely converted to Diesel power. Fifty-three American railroads now operate exclusively with Diesel-electric locomotives; thirteen others use Diesel-powered locomotives exclusively, with steam locomotives only as reserve or emergency units. With well over 85 per cent of all locomotives now at work on American railroads powered by Diesel-electric units, the total savings in fuel and maintenance are estimated as $600,000,000 yearly—with yearly savings of $37,000 a unit by no means exceptional.

Diesel power has greatly reduced the number of locomotives required for everyday rail operations. On the Erie Railroad, for example, 472 Diesel-powered locomotives now do the work of 1,545 steam locomotives. In Canada, where more than 500 are in service, Diesel-powered locomotives draw all the principal trains on the 2,881-mile run between Montreal and Vancouver; they operate successfully between Skagway, Alaska, and White-horse, Canada, even in the snow-bound mountains of Yukon Territory.

Other uses push forward. Of the estimated two million Diesel engines now in the United States at least one-fifth are currently being used in tractors and building contractors' equipment. For this group, about thirty thousand new units are being produced every year. The list of uses includes pumps and generator sets, heavy duty trucks, hoists, cranes, graders, cement mixers, bull-dozers, and earth-moving machines. Our vast resources of high-way and road-building machinery are now almost all Diesel powered. Much the same is true of excavating and heavy build-ing machinery. More and more of the nation's half-million oil wells use Diesel power for pipelining crude oil to refineries and petroleum products to principal distribution centers.

In 1946 about 10 per cent of American-built buses were powered by Diesel engines. By 1952 nearly 40 per cent were so powered. Practically all now have Diesel engines. More than 75 per cent of all heavy-duty trucks now in service, or over 135,000, have been converted to Diesels. The great saving in fuel is far exceeded by savings in maintenance costs. One major trucking firm overhauls its Diesel-powered trailer units annually or after 180,000 miles of road service. In Europe, where gasoline prices are unvaryingly high, the Diesel-engine automobile gains in popularity.

The United States' production of Diesel marine engines averages about 12,000 units a year. The largest individual user is the United States Navy—with some 40,000,000 horsepower currently in use as compared with about 140,000 horsepower following World War I. Fireboats, river barges, and ocean-going craft are being powered more and more by Diesel engines. Mining, municipal power plants, airports, and air bases are all showing impressive gains in their use of Diesel power. Electric generating plants alone, numbering more than 2,500, account for more than two million of the Diesel horsepower now in use.

The United States continues to lead the world in both use and manufacture of Diesel engines, but the multiplying descendants of Rudolf Diesel's "black mistress" are now serving people globally.

Significantly, more and more nations gauge their industrial growth in terms of the output of Diesel engines. Soviet Russia is no exception. In *Information U.S.S.R.*, Volume I of *Countries of the World, Information Series* edited by Robert Maxwell, one finds the figures indicating the multiplying use of Diesel power in Soviet Russia. Beginning with an estimated 35,000 horsepower in 1913 (almost certainly an underestimate) and 39,000 horsepower in 1928, the total climbed to 96,000 in 1932, 260,000 in 1937, 3,225,000 in 1950, 4,005,000 in 1955, and 4,403,000 in 1956—the latest tabulation.

The Kremlin 1965 goal for Diesel and electric power in rail transportation is 85 to 87 per cent of the total. In 1940 the comparative uses in railway traction was 2 per cent electric and 0.2 per cent Diesel power. By 1958 these ratios were reported as 15.1 per cent electric and 11.3 Diesel; in 1960, 21.8 per cent electric and 21.4 Diesel. As 1960 ended, the Kremlin reported a total of 31,600 rail kilometers converted to Diesel and/or electric traction. The target for 1965 has been listed as 100,000 kilometers.

Not surprisingly, the Soviets report that Diesel-powered locomotives are most effective and economical on distance runs and across great arid regions. This is a consensus of the findings in practically all operations of long-line railroads. The electrification of long lines bearing light to moderate freight traffic is rarely justified because of the cost of installation and operation. On shorter lines through densely populated areas, electrification is frequently more economical than Diesel power particularly when fuel must be imported at high costs and without reliable sources. But for long hauls and continuous runs (such as for transporting ore, sugar cane, and similar raw products), Diesel power holds and increases its advantages.

One of the most interesting examples of Diesel power on rails is being provided by West Germany. Two factors here work against the extension of long-proved Diesel power. One is the relatively high cost of fuel. The other is the change from the predominantly east-west traffic to predominantly north-south traffic as a result of the political division of Germany. This shift involves directing the heaviest rail traffic across three minor mountain ranges, which gives advantage to electrification. Even so, Diesel power keeps gaining strongly.

The number of Diesel locomotives rose from 1,089 in 1961 to 1,406 in 1962. In 1962, 6,259 steam locomotives were in service, a decrease of 505 from 1961; and there were 1,310 electric units, an increase of 182 over the 1961 figures. The latest *Bundesbahn* estimate of Diesel locomotives at work is 1,397 units. These operate on 55 per cent of the present rail mileage and provide some 70 per cent of the switch engines, with 400 large or medium engines assigned to freight or light passenger trains.

The German story of Diesel engines continues to move back and forth like a hard-driven piston. The close of the first World War brought a long pause in Germany's industrial progress. The long lapse continued from 1919 through 1924. The progress of the Diesel engine shared a similar fate, for by then it had grown to be a yardstick of Germany's foreign trade, which had virtually collapsed.

But M.A.N.—*Maschinenfabrik* Augsburg-Nürnberg—whose

basic story has already been told, kept resolutely at work. In 1920 the firm developed and patented an improved air-stream cleaning system, called *Umkehr-Schlitzspühlverfahren,* and a *"Zuganker"* construction, which made the assembly considerably easier without loss of stability.

During the previous year, the firm's developmental work had concentrated on a new six-cylinder, four-stroke-cycle engine with an extremely low work speed—100 rpm. Tests were undertaken with a 2,000-horsepower load. The first model engine was finished in 1921; after three years it was first employed in a shipyard.

By 1924, M.A.N. had other types of Diesel engines for both marine and stationary uses, including a single-acting, two-stroke-cycle engine with a bore of 31.5 inches and a stroke of 41.34 inches. A somewhat smaller 1,500-horsepower unit followed. Encouraged by the success of the new models, the company undertook the manufacture of double-acting, two-stroke-cycle engines. These found ready acceptance by shipbuilders. Thus the lighter double-acting, two-stroke-cycle engine began proving its value, which promptly led to building much bigger power units, with outputs between 2,500 and 15,000 horsepower.

Beginning in 1920, M.A.N. developed and tested a Diesel engine suitable for powering trucks. Based on the "Otto engine," this pioneer truck engine operated without an air compressor but with direct fuel injection; it developed forty horsepower at 900 rpm. An improved model, developing forty-five horsepower at 1,500 rpm, was demonstrated at the 1924 *Automobil-Ausstellung* in Berlin. The excitement caused by this very special exhibit at the big automobile show, which included a Diesel panel truck by Benz & Cie. of Mannheim, is still remembered.

By 1930 all trucks and buses manufactured by M.A.N. were Diesel powered. In 1934, a M.A.N. Diesel-engine truck won first prize for the 3,000 kilometer (1,863 mile) reliability run through Russia. M.A.N. next developed two sizes of four-cylinder automotive engines (twenty-five and fifty horsepower, at 1,500 rpm) for powering tractors. The new models were water-cooled. They were supplemented with a special-purpose 16-cylinder air-cooled engine of seven hundred horsepower and nine hundred horsepower with superchargers. M.A.N. also produced a variety of stationary engines, ranging from 30 to 300 horsepower a unit, and for railroad use a V-type engine of 450 horsepower and a twelve-cylinder 275-horsepower "Boxer" engine.

During 1927, the company set out to improve all its small engines. In the course of that year Russia and Spain urged the company to resume manufacturing Diesel engines for submarines. M.A.N. complied by building two types, the larger one for Soviet use. The larger unit was supercharged by an independent, mechanically driven blower. The smaller engine of 15.75-inch bore and 18.11-inch stroke set an impressive record for fuel efficiency—160 grams per horsepower hour.

In 1924, the first Diesel rail car at the German *Reichsbahn* was put in work. It was powered by a six-cylinder, four-stroke cycle, 75-horsepower engine, turning at 1,150 rpm without a high-pressure fueling system. A year later a large Diesel-electric freight locomotive, powered by a six-cylinder, four-stroke-cycle engine of 1,000 to 1,200 horsepower at 450 rpm with air compressor, was delivered to the *Reichsbahn*. In 1935, M.A.N. built and delivered an eight-cylinder Diesel engine of 1,400 horsepower (at 700 rpm) to power a passenger train locomotive. Three years later the first of a four-unit speed rail car was equipped with an eight-cylinder, 1,300-horsepower (at 700 rpm) engine.

Beginning in 1927, the Diesel engine factory began proving the power of its product as a builder of international trade by converting the power plant at Maria Elena, Chile, to Diesel engines. The procedure began with single-acting, six-cylinder, four-stroke-cycle Diesel engines developing 10,500 horsepower (at 150 rpm) and progressed to double-acting ten-cylinder, two-stroke-cycle engines of 11,700 horsepower (at 214 rpm).

During 1938–39, M.A.N. again improved its production line engines and added a 24-cylinder, V-type engine unit with an eventual horsepower output raised to twenty thousand. In 1939, the firm developed a double-acting, two-stroke, eight-cylinder Diesel engine of twelve thousand horsepower (at 245 rpm), which was presently put to active use by the German Navy.

The most impressive attainment was the continued reduction of weight and bulk of the rational heat engines. A 24-cylinder, V-shaped Diesel engine of sixteen thousand horsepower (at 450 rpm) including accessories weighed nine kilograms (19.8 pounds) per horsepower, less than one twenty-second part of the weight of the original Diesel engine of 1897, which weighed about two hundred kilograms (440 pounds) per horsepower. More recently weights of various types of Diesel engines have dwindled to two kilograms (4.4 pounds) per horsepower.

On October 4, 1937, M.A.N. placed an unpretentious plaque at the site of the old Diesel engine laboratory on its grounds. During February, 1944, the great factory was almost completely destroyed by Allied air attacks. But M.A.N. rebuilt. It remains one of the major builders of Diesel engines. The current output is well over 500,000 horsepower a year.

To some extent paralleling M.A.N.'s Diesel story is that of the Fried. Krupp *Werke.* In July, 1906, Ludwig Noé, who had served as an engineer in the Diesel office, took over as chief engineer for Krupp's Germania Shipyards. During 1907, the *Germaniawerft* received an order to manufacture several submarines for the Russian government. For powering these, Noé and his colleagues designed and tested a double-acting, four-stroke-cycle, four-cylinder Diesel engine, with double piston rods to produce two hundred horsepower at 450 rpm. This first direct-reversible engine weighed thirty-three kilograms (72.6 pounds) per horsepower and used about 180 grams (six ounces) of fuel oil per horsepower hour—with an entirely invisible exhaust.

Beginning in 1908, the *Germaniawerft* also built a series of stationary Diesel engines of vertical construction operating on a four-stroke-cycle. It built marine engines of steel and brass, using cast iron for the stationary units, for weight remained a crucial problem. A 450-horsepower, six-cylinder, four-stroke-cycle marine engine presently achieved a saving of forty kilograms (88 pounds) per horsepower. Krupp men also adopted a procedure for casting two cylinders *"en bloc."* This was applied in making two-, four-, and six-cylinder units for use on barges and tugboats, as well as for stationary purposes. Two 120-horsepower engines, operating at 400 rpm, were built for a side-wheeler to ply the shallow waters of the western Siberia river Ob and its tributaries. The battery of two Diesel engines weighed 11,000 kilograms— 24,200 pounds.

While developing the four-stroke, high-speed engines, the Germania yards also developed low- and high-speed, single- and double-acting two-stroke cycle engines for tankers of the German-American Petroleum Corporation. A subsequent 1,800-horsepower Diesel unit proved adequate for a 15,000-ton tanker, but engines with power output up to 3,000 horsepower were developed for larger vessels. A four-cylinder, 140-horsepower unit (at 250 rpm), weighing from 220 to 264 pounds per horsepower, was developed for use on tugboats.

But for warships and other vessels where weight is an especially important factor, the Germania yards managed to produce Diesel engines which served effectively with weights averaging around seventy-seven pounds per horsepower. The next step was designing and producing a still lighter 1,200 horsepower engine (at 450 rpm) for installation in submarines, with similar but much larger (4,500 horsepower) engine for use in torpedo boats.

By 1911, Diesel power for submarines was the biggest challenge. Directly before the outbreak of the first World War, the Krupp-owned shipyard powered the first commercial cargo submarine, *Deutschland,* with an enclosed, vertical six-cylinder, four-stroke-cycle Diesel engine which developed 450 horsepower (at 400 rpm). This *"Handels-U-Boot"* called at Norfolk, Virginia, on May 3, 1916, for a cargo of critical war supplies. Two months later it visited New London, Connecticut. Krupp men also developed a master ship's Diesel engine capable of developing twelve thousand horsepower, but were unable to put it to use.

After the close of the war, Krupp's yards and plants returned to building the more basic two-stroke-cycle Diesel engines in units up to eight cylinders and 4,500 horsepower, for installing in power plants. Then a more comprehensive program of Diesel engine building was resumed. It included large, low-speed, two-stroke-cycle engines for ship propulsion, stationary and auxiliary marine engines, and four-stroke-cycle engines for smaller boats, trucks, and locomotives. In 1935, work began in the *Krupp-Südwerke* at Essen on two-stroke-cycle Diesel engines which were built in units of three, four, and six cylinders, developing 110, 145, and 210 horsepower respectively, all with exhaust turbochargers. This was one of the most effective, versatile programs of developing Diesel power.

During the second World War, the Krupp works concentrated on building Diesel engines for submarines for the German Navy. The Fried. Krupp *Werke* now builds Diesel-powered heavy trucks and utility vehicles.

Another chapter of German progress in Diesel power tells of the fortunes and misfortunes of *Motorenfabrik* Deutz. In 1930, the Deutz works combined with the *Maschinenbauanstalt* Humboldt, Germany's oldest manufacturer of mining machinery, to form *Humboldt-Deutz-Motoren,* A.G.

Six years earlier Deutz began test production of Diesel-powered trucks which *Motorenfabrik* Oberursel built. Within two years the effort had developed a high-speed Diesel engine suited to automotive use. In 1930, Deutz absorbed Oberursel and in 1936 merged with the truck manufacturing firm, Magirus of Ulm, which specialized in building fire-fighting equipment. The union, which merged expert skills and experiences, produced trucks equipped with water-cooled Diesel engines of four, six, or eight cylinders developing 120 to 180 horsepower. Rotos or Büchi superchargers were optional equipment.

Before the outbreak of the second World War several German manufacturers were producing air-cooled Diesel engines. In 1944 the Ulm works began series production of the "air coolers." During the war, German output of air-cooled Diesel units increased. The trend was restored following the rebuilding or repair of war-blasted factories.

The pioneering Deutz firm is now the Klöckner-Humboldt-Deutz, A.G. It has continued to build air-cooled automotive Diesel engines ranging in output from 10 to 250 horsepower. Since the "pre-combustion chamber" type of engine developed too much heat for air cooling, the engine is now built with a "turbulence chamber" instead of the "pre-combustion."

The Deutz works next (in 1949) began manufacturing air-cooled Diesel engines for powering tractors. These were immediately successful. By 1954 their sales had risen well above the 130,000 mark, with engines capable of speeds up to 2,800 rpm. Meanwhile, the firm's Magirus Division began producing two basic truck models: one powered with a 90-horsepower Diesel engine, the other with a V-type engine, eight-cylinder, 175-horsepower for trucks, buses and various other utility vehicles.

Present production of the Deutz and Kalk works includes horizontal Diesel engines; three-cylinder verticals; two-stroke- and four-stroke-cycle engines of from one hundred to six hundred horsepower; an eight-cylinder, four-stroke-cycle unit of up to two thousand horsepower; and a high speed, two-stroke-cycle engine for marine use.

The locomotive division now produces Diesel engines for underground work, narrow-gauge railroads, standard-gauge switch engines, air- or water-cooled engines for powering generators, water pumps, air compressors, various kinds of farming machinery, and auxiliary marine engines. In terms of total units produced, the Klöckner-Humboldt-Deutz of Cologne is probably the world's largest Diesel engine manufacturer. More than 750,000 air-cooled Deutz-Diesel engines, totaling some forty million horsepower for powering vehicles, bear its trademark.

In addition to M.A.N., Krupp, and Deutz, there are other old-line German Diesel engine makers, some of which entered the field prior to 1900. L. A. Riedlinger, *Maschinen und Bronzewarenfabrik* of Augsburg was one example until 1908 when M.A.N. absorbed it. Another very early contender was H.

Paucksch of Landsberg on the Warthe which helped pioneer the engine in the 1890's.

As already noted, Germany's great automobile makers, Benz & Cie. of Mannheim, helped develop the Diesel engine for automotive use. This pioneering began in 1908 when Benz employed Rudolf Diesel's admirer, Prosper L'Orange, to direct the developmental work. L'Orange's patent for a pre-combustion chamber, issued in 1909, opened the way for the small, high-speed automotive engine.

The Benz company also built Diesel marine engines, including those built into the motorship *Hermann Krabb,* which early in 1913 made her maiden voyage to South America.

By 1912 the Daimler *Motorengesellschaft,* presently to merge with Benz, had also developed and placed in production marine Diesel engines of sixty and one hundred horsepower (at 750 to 800 rpm). In 1917 the company tested a six-cylinder, 1,700-horsepower (at 380 rpm) Diesel engine for powering submarines. In 1924, Benz first exhibited farm machinery powered by Diesel engines at the Königsberg Agricultural Fair.

The merger of Benz and Daimler in 1926 intensified work on Diesel engines for automobiles. The firm promptly set out to produce and market a two-ton Diesel-powered truck. It began producing larger units for stationary use and also for marine and locomotive installation. The airships *Hindenburg* and *Graf Zeppelin* were both powered with Daimler-Benz 1,200-horsepower Diesel engines.

During 1936, Daimler-Benz began producing Diesel-powered passenger automobiles. The four-cylinder, 2.6 liter engine cars attained a maximum speed of sixty-five miles an hour with a fuel consumption of 2.6 gallons per one hundred miles. The company now builds a wide range of automotive Diesel engines, stationary, marine, and rail units, ranging from one to twenty cylinders and 13 to 2,500 horsepower.

Several other manufacturers in Germany continue to produce trucks, heavy duty vehicles, and passenger autos powered by Diesel engines. Many German Diesel engine makers, particularly

Krupp, M.A.N., Daimler-Benz, Maybach, Voith, Krauss-Muffei, Esslingen *Werke*, Jung, M.A.K. and Henschel, have helped in the development of the new high-speed engine.

Through the years, the Swiss firm which provided Rudolf Diesel his first factory job has proved one of the all-time great champions of Diesel power. Sulzer contributions to Diesel power are legion. We have already noted at least a dozen. In 1914, Sulzer Brothers delivered to the Prussian and Saxon State Railways five rail cars of the same general type as the McKeen car which R. Diesel had traveled from St. Louis, Missouri, to Fayetteville, Arkansas, to investigate. The McKeen car was patented in 1909; the "independent unit" Sulzer car was patented almost exactly five years later. The Sulzer car was a superior unit primarily because of its Diesel engine, a four-stroke, six-cylinder, V-type of 200 horsepower (operating at 400 rpm).

When the first world war ended, Sulzers renewed their work with the Diesel engine by producing railroad engines which could be started with storage batteries instead of compressed air. These were followed by a "double row," V-type engine for railroad locomotives. Typical of that construction was the series of 4,400-horsepower locomotives of 1938, using two Diesel engines of 2,200-horsepower each, with two parallel rows of six vertical cylinders acting on two parallel crankshafts. Sulzers also built the "verticals" in single rows of six or eight cylinders.

By 1924 almost one-third of the total motorship tonnage then in use was being equipped with Sulzer Brothers Diesel engines. In that year Glasgow's Fairfield Shipbuilding and Engineering Company completed and launched the liner *Aorangi*, "Queen of the Southern Seas." Powered with four six-cylinder, Sulzer-designed propulsion engines, totaling thirteen thousand horsepower, the 23,000-ton liner, carrying thirteen hundred passengers and a ship's company of more than three hundred, maintained a speed of eighteen knots on its regular run between Vancouver, Auckland, and Sydney.

By 1938, Sulzer Diesel engines had scored still another record for power and speed in passenger ship service. The champion then was the Netherlands' 24,700-ton M. S. *Oranje*. Equipped with three twelve-cylinder, single-acting, direct-reversible, two-stroke-cycle Sulzer engines (12,500 horsepower at 145 rpm), *Oranje* reached a speed of 26.3 knots.

At the present time Sulzer Brothers factories are located in Winterthur, Oberwinterthur, Soleure, and Bülach (Switzerland), and its branch offices are located in major cities throughout the world. In its first half-century of manufacturing, Sulzer Brothers produced Diesel engines of more than ten million horsepower, not including the many license holders which together produced even more. Currently the Diesel Department of Sulzer Brothers produces low- and high-speed four-stroke-cycle engines; single- and double-acting two-stroke-cycle engines for stationary plants; two-stroke-cycle marine engines for passenger and cargo ships; marine auxiliary engines for powering pumps and generators; Diesel engines for railcars and locomotives; auxiliary compressors for stationary and marine plants; and exhaust-gas heat-recovery equipment.

Another early Swiss manufacturer of Diesel engines was the Adolph Saurer Company of Arbon. Saurer's contributions have been in developing engines for automobiles and for powering textile machinery manufactured by the concern. During 1908, Rudolf Diesel worked directly with Saurer engineers to convert one of its gasoline truck engines to the Diesel principle. The success of this venture in developing a four-cylinder, medium-speed, 25-horsepower engine for truck use caused the company to undertake extensive manufacture of the truck engines. By 1928, Saurer was producing usable units in substantial numbers. The earlier models had an Acro air chamber, which was succeeded by a crossflow water-cooling system. Next came direct fuel injection and a dual-turbulence air-flow system. By 1935 the company began making small, high-speed Diesel engines for light delivery trucks and large passenger automobiles. By successive stages the ingenious firm, which had been ably led by Hippolyt Saurer, developed an exhaust-gas turbo-charging system for

Diesel engines. By 1940, Saurer had begun manufacturing larger stationary engines, and in 1942 it began building 700-horsepower Diesel units for locomotives.

At the present time the Adolph Saurer Company builds a series of Diesel engines of four, six, eight, and twelve cylinders, developing from fifty-four to three hundred horsepower for both automotive and stationary uses.

The name of the pioneer Diesel engine concessionaire of Amsterdam *Nederlandsche Fabriek van Werktuigen en Spoorweg-Materiel* has been reduced to *Werkspoor*, N.V. (*Naamloose Venootschap* or corporation).

Directly after World War I, the firm designed a new type of double-acting, four-stroke-cycle Diesel engine. By 1924, twelve of the new engines had been completed and installed in tankers of 14,900 tons displacement. During 1926, *Werkspoor* developed a new supercharging system in which pre-compression of the combustion air was effected by the lower part of the working piston in the closed under-cylinder.

In 1932, *Werkspoor* engineers designed a new marine engine for ocean-going tugs. The direct fuel-injection engine had a greatly lowered crosshead, separately driven supercharging pumps, and a reductor with hydraulic couplings to transmit the three thousand horsepower developed by the two main engines which drive one propeller shaft.

At the present time *Werkspoor*, N.V. builds a single-acting, four-stroke-cycle Diesel engine of trunk-piston construction with mechanical fuel injection, forced lubrication, and Büchi supercharger. This engine is manufactured in four cylinder sizes and in units of as many as ten cylinders. Power outputs range from 150 to 2,100 horsepower. *Werkspoor* also builds two-stroke-cycle, crosshead engines with uniflow scavenging in three cylinder sizes and from four- to twelve-cylinder units, with output ranging from 1,800 to 9,600 horsepower. The Dutch firm manufactures, too, a high-speed, supercharged Diesel engine suitable for locomotives, rail cars, ships, and stationary power units. These engines develop from 475 to 3,000 horsepower at speeds up to fourteen hundred revolutions per minute.

In 1915, Mirrlees of Scotland and England contributed to the growth of Diesel power by developing an electrical-mechanical device for correlating automatically air-injection pressures with the prevailing engine load. This advance proved particularly helpful in multi-engine operations. Another important Mirrlees introduction was the floating piston pin. This feature is now in almost universal use. Prior to the innovation the usual practice was to lock the pin to the piston, a positioning which often caused piston seizure and distortion. The innovation permits axial and rotational freedom. Mirrlees engineers also developed an indicator for aligning crankshafts.

Mirrlees has pioneered other technical improvements destined to be of widespread benefit to the rational heat engine. By 1925 the company had perfected a two-stroke-cycle Diesel engine with a 22-inch diameter bore and a 30-inch stroke, developing three hundred horsepower per cylinder at 150 rpm. The pistons were oil-cooled, and the engine fitted with compressors operated by rocker arms. These were later replaced by compressor drives at the end of the crankshaft. This model helped produce the first widespread modernization of cities in South America.

During 1925, Mirrlees developed other enclosed Diesel engines with forced lubrication especially for marine use. During the mid-1930's the firm developed its "work-horse Diesel," building in line units of from three to eight cylinders and V-type units of twelve cylinders with individual cylinders developing sixty horsepower at 900 rpm.

Following World War II, Mirrlees, by then Mirrlees, Bickerton and Day, Ltd., joined the Brush Group thereby gaining close associations with firms in Australia, Belgium, Brazil, Canada, Ireland, Malaya, South Africa (Rhodesia), the United States, and Venezuela. The consolidated company continues as the oldest Diesel manufacturing company in Britain and at the present time is the Commonwealth's foremost producer of Diesel engines.

In 1949, with practically all its machine tools recast, Mirrlees set out to develop two basic types or "ranges" of Diesel engines. One, with a bore of 9.75 inches and a stroke of 10.5 inches, is made in lines of three to eight cylinders and V's of eight, twelve, and sixteen, with a range of 135 to 2,480 horsepower at 600 to 900 rpm. The more powerful range is an enclosed power plant featuring a cylinder of fifteen inches and a stroke of eighteen inches, with power outputs ranging from 183 to 3,060 horsepower at 200 to 450 rpm. These two basic types of Diesel engines have found a multitude of uses in marine auxiliary engines, rail traction units, marine propulsion, and stationary generating units for both land and sea uses.

Currently, at least thirty-seven different British firms are manufacturing Diesel engines for industrial, rail, and marine uses only. Among the too little honored British pioneers is Scott and Hodgson, Ltd., of Manchester, which built, in 1900, what was probably the first two-stroke Diesel engine ever produced. It was a horizontal, single-cylinder engine.

In 1938, British exports of Diesel engines for other than automotive use amounted to a respectable three million pounds sterling. The peak was reached in 1952 when Diesel engine exports amounted to well over thirty-five million pounds sterling. Subsequent foreign sales of Diesel engines have held up, though somewhat below the 1952 peak figure. Significantly, about three-fourths of British manufactured Diesel units go to less developed countries abroad. The same is rather generally true of other European producers.

Diesel history in Italy dates back to 1903, when engineer Franco Tosi of Legnano acquired a license to build Diesel engines in Italy. The Tosi firm, established in 1874, had gained recognition as a manufacturer of gas engines and steam engines.

Of all the talented Italian group, the Tosi company has most clearly shown the overall progress of the Diesel engine.

By 1914 the company had developed a "gigantic" unit developing 2,400 horsepower. In 1925 its engineers developed a mechanical fuel-injection system which served to eliminate the multi-stage compressor and reduced fuel consumption by at least 10 per cent. Four years later Tosi perfected a turbo-charged Diesel engine, the second of its type ever producd. By 1932 it had built and installed in a power plant two ten-cylinder engines of 8,000 horsepower each. Within another decade it developed and put in use a 6,400-horsepower, eight-cylinder Diesel engine for marine use. The firm now manufactures turbines, pumps, and a great range of two-stroke and four-stroke Diesel engines for stationary and marine purposes, in various sizes and in units of from four to sixteen cylinders.

Italy's FIAT (*Fabrica Italiana Automobili Torino*) first began work on Diesel engines in 1906 under direction of Lauro Bernardi. It produced a marine unit in 1907. Three years later Fiat delivered to the German government a 1,200-horsepower engine for submarine use. Currently Fiat makes Diesel engines for its various lines of automobiles, buses, and light and heavy trucks. Several other automobile makers of Italy are similarly engaged.

Although modest in volume, Scandinavian production of Diesel engines makes a brilliant story. By 1928 a Danish firm was successfully pioneering Diesel engines without the impediment of air compressors and with proved mechanical efficiency raised to a remarkable 82 per cent.

An even more basic contribution was Burmeister and Wain's remarkable success in reducing the weight and bulk of the engines without detriment to either their efficiency or durability. By 1930 they had reduced the weight of a single acting, four-stroke-cycle Diesel engine from an initial 404.8 pounds per developed horsepower to 134.2 pounds.

By 1937 when the new S.S. *Selandia* replaced its predecessor, the new ship of 8,500 tons was powered by Burmeister and Wain's Diesel engines developing 7,300 horsepower. The "old" *Selandia* also had been powered by Diesel units of comparable thrust, but in a mere sixteen years the space required by the Diesel power plant had been reduced to less than one-half and the weight had been reduced from about 240 pounds per developed horsepower to 136. Simultaneously, cruising speed was increased from eleven to sixteen knots an hour.

In 1914, Copenhagen's famed *Aktieselskabet* Burmeister and Wain's *Maskin- og Skibsbyggeri* began developing Diesel engines for submarines. Diesels for locomotives were put into manufacture in 1920. The company stays most famous, however, for its marine engines. More than a thousand ocean-going vessels, including tankers, fruit ships, general cargo ships, passenger liners, and motor yachts are currently powered by Diesel engines made by the Danish firm.

Another early Scandinavian manufacturer of Diesel engines is A/S Frichs of Aarhus, Denmark. Frichs completed its first Diesel engine in 1909; its first Diesel locomotive in 1925. Since then the veteran firm has concentrated on manufacturing various types of Diesel-powered locomotives and rail cars. It has also a new series of Diesel engines of medium speed (from 50 to 1,000 horsepower) for use as stationary power plants and for auxiliary marine use.

The power story of the Diesel engine goes on, on every continent, and in at least seventy-five nations. It is well begun, though many authorities are convinced it is just barely begun.

The Diesel Engine Aloft

THE DEVELOPMENT OF THE DIESEL ENGINE for aviation actually began in 1904, but the first successful Diesel-powered test flight was not undertaken until twenty-four years later. In 1902, Franz Lang, a German engineer, began experimenting with a small Diesel engine capable of producing high speeds.

In 1908, Prosper L'Orange perfected an injection system without air compression and with a pre-combustion chamber. While working at Benz & Cie. of Mannheim, in 1909, he received a German patent (No. 230,517) for his invention.

The following year, the airplane builder Hugo Junkers received a patent for his "all-wing" airplane. Junkers turned to the development of the Diesel engine for airplanes, when it was a radical concept. There is reason to believe that Junkers was attracted to the Diesel engine because its fuel was not highly flammable. The ignition system of the exposed gasoline aero-engine had caused considerable trouble in adverse weather and presented many vexing problems.

The first Diesel aviation engine built by Junkers was a four-cylinder engine, tested in 1913. A six-cylinder engine was completed the following year. In 1916, Junkers built a six-cylinder Diesel aviation engine of 500 horsepower at 2,400 rpm. It was not until ten years later that a new five-cylinder Diesel engine, developing 830 horsepower (at 1,200 rpm) was constructed by the Junkers *Flugzeug und Motorenbau Gesellschaft.*

This engine, redesigned as a six-cylinder unit, was installed in a Junkers airplane and test-flown. A flight from Dessau, the factory site, to Cologne was made on February 4, 1929. Two years later a 720-horsepower Diesel engine (at 1,700 rpm) was placed in regular service by the *Deutsche Lufthansa*. This engine became the well-known Jumo (for Junkers *Motorenbau*) 205 engine. It was built until the beginning of World War II.

The power of the Jumo 205 was soon raised to 880 horsepower (at 3,000 rpm) and to 1,200 horsepower with the addition of an exhaust-driven supercharger. The six-cylinder, two-stroke-cycle engines had two crankshafts connected by a gear train. The opposed pistons made it virtually a twelve-cylinder engine, as the space between the pistons acted as a combustion chamber. The cylinder block and pistons were made of aluminum alloy, and the water-cooled engine weighed about 1.4 pounds per horsepower.

Mainly through the efforts of Franz Lang, a radial-Diesel airplane engine was developed. A patent was granted to Lang in 1926 and another in 1930. The resulting Diesel engine, which developed 650 horsepower (at 2,200 rpm), was built in 1938 by the famed *Bayerische Motoren Werke,* manufacturers of motorcycles and automobiles. It was the nine-cylinder BMW–Lanova.

In 1934 a further development of their Diesel engines for trucks and buses led the Benz company, after the merger with the Daimler *Motoren Werke* (in 1927), to build Diesel airship engines. The four-stroke-cycle 602 had sixteen cylinders of 175-millimeter bore and 230-millimeter stroke, arranged in a V-shape. The engine developed 1,320 horsepower at 1,650 rpm and weighed 3.3 pounds per horsepower. At a continuous cruising speed, the engine produced 800 horsepower. Installed in the airships *Hindenburg* and *Graf Zeppelin*, the four Daimler-Benz engines provided the dirigibles a flight range of 8,700 miles. The ideal combination of Diesel engines for propulsion and helium gas for lifting was, unfortunately, never achieved to make this means of transportation an ideal one. The *Hindenburg,* filled with highly combustible hydrogen gas, burned upon landing at Lakehurst, New Jersey, in 1937.

The Junkers Ju 86 airliners were powered by two 600-horse-power Diesel engines, and the Dornier Do 26 flying boat was equipped with four Jumo 700-horsepower engines. The German airforce built large numbers of Junkers two-engine fighter bomb-ers, Dornier seaplanes, and Blohm and Voss 138 patrol flying boats, powered by three Diesel engines of six hundred horse-power each.

A two-engine Junkers mailplane made a 3,600-mile flight from Germany to Bathurst, West Africa, in 1936. Regular mail service across the South Atlantic was carried out in 1937 by Dornier Diesel-powered seaplanes and later by four-engine Blohm and Voss planes. Air launched by catapult from the *Westfalen* in the English Channel, a Dornier flew the 5,125 miles to Brazil in forty-five hours and ten minutes. In 1936, eight survey trips were made between New York and the Azores; in 1937, fourteen flights; and in the following year, twenty-six flights were made— all in Diesel-engine seaplanes built by either Dornier or Blohm and Voss.

In France the engineer Pierre Clerget exhibited his 100-horsepower (1,800 rpm) nine-cylinder, air-cooled, radial-Diesel aircraft engine in Paris in 1930. It was test flown on September 30, 1929, at a moraine (glacial deposit) at Villacoublay. Further development was also carried on by the Hispano-Suiza works, but later France's Air Ministry took over the entire project.

In 1937, a Potez observation plane, powered by a radial-Diesel engine of 940 horsepower (at 2,400 rpm) reached an altitude of 25,114 feet. Two years later a large sixteen-cylinder, water-cooled, turbo-charged, double-star Diesel engine of 2,000 horse-power was bench-tested. It was called the *"Transatlantique."*

Other French aircraft engine developments included the water-cooled radial Salmson Diesel engine of eighteen cylinders which developed six hundred horsepower (at 1,700 rpm). Three different types were designed by Jalbert-Loire, a four-cylinder Diesel engine of 160 horsepower, a six-cylinder engine of 235 horsepower, and a sixteen-cylinder engine of 600 horsepower (at 2,400 rpm). France also had to its aeronautics credit a nine-cylinder Diesel engine built by Lorraine of 200 horsepower (at 1,500 rpm); an eight-cylinder, air-cooled Botali of 118 horsepower (at 2,000 rpm); and a seven-cylinder, air-cooled Delafontaine Diesel engine of 400 horsepower. However, none of the last named engines was ever flight-tested. Diesel aircraft engines of 600 horsepower (at 2,200 rpm), built by a license holder of the Junkers and based on that design, were test-flown, but were not further developed in France.

In Great Britain, the Bristol Aeroplane Company began work, under the direction of Roy Fedden, on the Phoenix nine-cylinder, air-cooled, radial-Diesel engine which developed 430 horsepower (at 2,000 rpm). Test flights were made in 1933 in a Westland Wapiti observation plane. Equipped with a similar engine with a supercharger, an airplane piloted by H. J. Penrose reached an altitude of 27,453 feet on May 11, 1934, setting a world's altitude record for Diesel-engine airplanes.

In 1930 another Diesel aircraft engine was constructed by the William Beardmore and Company, Ltd. Five eight-cylinder, steam-cooled Diesel engines of 585 horsepower (at 900 rpm), weighing 7.8 pounds per horsepower, were installed in the Airship R-101. This dirigible was promptly destroyed in a severe storm over India. The Beardmore company also designed a horizontally opposed piston engine of twelve cylinders which developed 500 horsepower (at 1,750 rpm), but the construction of it was never completed.

Britain's D. Napier and Son, Ltd., built under license the Junkers engines which developed 720 horsepower (at 1,700 rpm); these were test-flown. However, the Cutlass engines of 535 horsepower never got into production. Two regular Rolls Royce gasoline engines were converted to Diesel operation; their output was limited—only 480 horsepower (at 1,900 rpm).

In 1953 the Napier Company built and tested a unique Diesel-plus-turbine aero-engine. The *Nomad* compound-Diesel engine had twelve horizontally opposed cylinders of 6-inch bore and

7.375-inch stroke, operating on the two-stroke-cycle at 2,050 rpm. It developed 4,095 horsepower at take-off. The compression ratio was 31.5 to 1. The cylinders were without poppet or sleeve valves. A twelve-stage axial-flow compressor was mounted below the Diesel unit and driven by a three-stage gas turbine mounted at the rear. The turbine, in turn, was fed by exhaust gases from the Diesel cylinders. The engine weighed 3,580 pounds—only 0.93 pounds per horsepower. Although heavier than a pure turbojet power unit, this compound engine showed extremely low fuel consumption in the exhaustive testing experiments at Luton. It operated equally satisfactorily on regular Diesel fuel, kerosene, or low quality gasoline.

The French engineer Coatalen had taken a Benz engine in 1926 and converted it to Diesel use. In 1929 he exhibited a six-cylinder Diesel-engine aircraft of his own design, which developed 104 horsepower at 1,500 rpm. It was built by the Sunbeam Motor Car Company, but it was never test-flown. When Coatalen later returned to France, he designed a six-cylinder aircraft engine at the Panhard works. This engine was followed in 1936 by a twelve-cylinder, water-cooled engine of 600 horsepower (at 2,200 rpm).

In the United States, Diesel aviation engines were developed by the Packard Motor Car Company, under the direction of Captain Lionel Woolson and the German engineer Herman Dohner. As chief aeronautical engineer of the company, Woolson had designed the regular Packard gasoline aircraft engines and those for the dirigible *Shenandoah*. In 1929 Woolson flew a Stinson monoplane from Detroit to Langley Field, Virginia, a distance of seven hundred miles, in six hours and forty minutes. The cost for the fuel was less than one cent a mile. The nine-cylinder radial-Diesel engine developed 225 horsepower (at 1,950 rpm) and weighed 510 pounds. At the Detroit Air Show, six airplanes and a Ford tri-motor eleven-passenger airliner were equipped with these engines. However, after Woolson was killed in a crash when his small plane was lost during a storm near Attica, New York, early in 1930, the Packard factory discontinued the production of Diesel aircraft engines. Even so, the Collier Trophy was awarded to the Packard Motor Car Company for the most important contribution to aviation in 1931.

In 1930, the Austrian engineer F. A. Thaheld designed a radial aircraft engine for the Guiberson Diesel Engine Company of Dallas, Texas. The following year this engine was installed in a Waco biplane and test-flown. Colonel Arthur Goebel reached an altitude of 21,686 feet in November, 1931. The engine developed 185 horsepower (at 1,925 rpm). In 1934, two larger engines of 253 horsepower at 2,100 rpm were tested, and in 1940 an improved type, developing 310 horsepower (at 2,150 rpm) and weighing 2.1 pounds per horsepower, was built and tested under government supervision. It was later used in armored tanks.

Other American Diesel aviation engines built, but not flight-tested, were the two-cylinder, 75-horsepower, water-cooled Attendu airship engines of 1925; the 400-horsepower, seven-cylinder, radial, air-cooled engines of 1932; and the Deschamps 1,200-horsepower, twelve-cylinder Diesel engines with two banks of inverted liquid-cooled cylinders, built in 1934.

A license to build Packard aircraft engines was acquired by the Walter Company of Czechoslovakia, but no engines were built. However, the National Arms Factory designed and built a nine-cylinder two-stroke-cycle, radial-Diesel engine in 1930. It developed 260 horsepower (at 1,560 rpm) and was widely used in training planes.

In Italy the large Fiat works at Turin built, in 1930, a six-in-line-cylinder, water-cooled Diesel engine of 220 horsepower (at 1,700 rpm) and weighing four pounds per horsepower. An earlier design of Italian origin, the Garuffa, had been shown at the Paris Aero Show in 1921.

More than sixty years ago Rudolf Diesel had sought to develop a new kind of engine. He could never have anticipated the practical uses to which it has been put. Aeronautics, like every other field in which the engine has been used, has been changed and advanced because of the Diesel engine.

Finale

At Broken Hill, Australia, a gigantic Swiss-made Diesel engine of 47,050 horsepower provides the area's electricity. Diesel-powered rail cars, built in the United States, speed across the searing deserts of the Middle East. Diesel engines built in the United Kingdom power ore dredges in Malayan tin mines. Diesel engines from the United States power irrigation pumps in Israel. Fishermen use Diesel-engine boats built in Scandinavia to troll fish-rich waters for the benefit of hungry people. Belgian-built Diesel engines power mining machinery which takes uranium ore from the Congo. Canadian-manufactured Diesel-electric locomotives pull trains on the East Bengal Railway.

Mining machinery in South Africa is powered by Diesel engines made in West Germany. Fast Diesel-powered trains, made in Switzerland, skim over the narrow-gauge rails of the Royal State Railway in Siam. The Holy City of Mecca is lighted by stationary Diesel units built in England. Diesel power helps to mine gold in Australia. In Japan a new factory builds Diesel engines producing one horsepower for use in under-developed countries and on small Japanese farms and local road-building.

And so the story goes and does not end. Diesel engines circle the earth and work wherever people seek an improved standard of living in every populated land. Diesel engines help to bring the family of man closer together. Rudolf Diesel would have liked that.

Chapter Bibliographies

Published sources consulted for individual chapters are listed below. Complete publication information is not repeated for items listed in the Diesel Bibliography following this section.

CHAPTER ONE

Chalkley, A. P. *Rudolf Diesel.*

Codrington, George W. *Shadows of Two Great Leaders—Rudolf Diesel and Alexander Winton.*

Crew, Henry. "The Tragedy of Rudolf Diesel," *Scientific American,* December, 1940.

Diesel, Eugen. *Germany and the Germans.*

———. *Jahrhundertwende.*

———. *Ringen um Europa.*

Holmgren, E. J. "Rudolf Diesel, 1858–1913," *Nature,* 1958.

Kennedy, John B. "The Cavalcade of Diesel," *Universal Engineer,* January, 1937, pp. 23–47.

CHAPTER TWO

Chalkley. *Rudolf Diesel.*

Diesel, Eugen. "Aus der Kindheit des Dieselmotors," *Westermanns Monatshefte,* Vol. CXLIII (1928), 511–14.

———. *Diesel.*

———. *Die Söhne Fortunas.*

Forrest, A. B. G. *Great Britain and the Industrial Awakening.* London, Longmans, Green, 1903.

Mason, K. Tolman. *The Franco-Prussian War.* London, The Macmillan Company, 1893.

285

Powers, E. C. "Rudolf Diesel," *Marine Review*, Vol. LVIII (1928), 91–132.

CHAPTER THREE

Holmgren. "Rudolf Diesel, 1858–1913," *Nature*, 1958.
The New York Times, June 3, 1912.
Powers. "Rudolf Diesel," *Marine Review*, Vol. LVIII (1928), 91–132.
Reminiscences of E. P. Shrienberg. London, Unwin Hill, Ltd., 1938.
Schnauffer, Kurt. "Die Erfindung Rudolf Diesels—Triumph einer Theorie," *VDF Zeitschrift* (1958).
Sulzer Brothers. *Historical Retrospect and Technical Developments.*

CHAPTER FOUR

Chalkley. *Rudolf Diesel.*
Diesel, Rudolf. "Der Mechanische Wirkungsgrad und die undizierte Leistung der Gasmaschine," *Zeitschrift des Vereins Deutsches Ingenieure*, Vol. XLIX (1905), 449.
Garfield, James S. *The Story of Patents.* New York, Appleton, 1940.
Holmgren. "Rudolf Diesel, 1858–1913," *Nature*, 1958.
Kilgew, R. E. *A Story of Mechanical Refrigeration.* London, Unwin Hill, Ltd., 1939.
Lossow, P. von, *et al. Neuere Literatur über den Diesel-Motor.*

CHAPTER FIVE

Crew. "The Tragedy of Rudolf Diesel," *Scientific American*, December, 1940.
Diesel, Eugen. *Diesel.*
———. *Das Gefährliche Jahrhundert.*
———. *Philosophie am Steuer.*
The Diesel Motor.
Reminiscences of E. P. Shrienberg.
Sulzer Brothers. *Historical Retrospect and Technical Developments.*

CHAPTER SIX

Diesel, Eugen. *Jahrhundertwende.*
Diesel, Rudolf. *Antwort auf Emil Capitaine's Kritik des Dieselmotors.*
———. "The Diesel Engine," *Progressive Age*, Vol. XVII (1899).
———. *Diesel's Rational Heat Engine.*
———. "Der Heutige Stand der Wärmekraftmaschinen und die Frage der Flüssigen Brennstoffe, unter besonderer Berücksichti-

gung des Diesel-Motors," *Zeitschrift des Vereins Deutscher Ingenieure.* Vol. XLVII (1903), 1366–75. (Cited hereafter as "Der Heutige Stand . . . .")

———. *Theory and Construction of a Rational Heat Motor.*

Kraus, Wolfgang. *Rudolf Diesel.*

Kurzel-Runtscheiner von, Erich. *Der Dieselmotor.*

Pachtner, Fritz. *Patent 67207.*

CHAPTER SEVEN

Diesel, Rudolf. "Der Mechanische Wirkungsgrad und die undizierte Leistung der Gasmaschine," *Zeitschrift des Vereins Deutscher Ingenieure,* Vol. XLIX (1905), 449.

———. *The Present Status of the Diesel Engine in Europe and a Few Reminiscences of the Pioneer Work in America.* (Cited hereafter as *The Present Status of the Diesel Engine* . . . .)

Stead, H. V. "Forty Years Onward," *Gas and Oil Power,* Vol. XXXII (1937).

CHAPTER EIGHT

Chalkley, A. P. *Diesel Engines for Land and Marine Work.*

Diesel, Eugen. "Aus der Kindheit des Dieselmotors," *Westermanns Monatshefte,* Vol. CXLIII (1928), 511–14.

Diesel, Rudolf. "The Diesel Engine," *Progressive Age.*

———. *Theory and Construction of a Rational Heat Motor.*

Diesel Engine Users Association of London. *No. S–100,* "The Development of the Multicylinder Horizontal Oil Engine."

———. *No. S–91,* "Modern Engineering Cast Irons and Properties."

Ensslein, Max. *Explosionsmotor und Verbrennungsmotor.*

Kennedy. "The Cavalcade of Diesel," *Universal Engineer,* January, 1937, pp. 23–47.

Maschinenfabrik Augsburg-Nürnberg. *Bilder, Dokumente und Erläuterungen zur Entstehung des Diesel-Motors unter besonderer Berücksichtigung der Entwicklung des Fahrzeug-Diesel-Motors.*

———. *Denkschrift zur Erinnerung an die Feier der Stadt Augsburg am Anlass der Enthüllung einer Rudolf Diesel Gedenktafel und an die Feierstunde der Maschinenfabrik Augsburg-Nürnberg A.G. anlässig der Eröffnung des M.A.N. Werkmuseums in Augsburg am 23. April, 1953.*

———. *Fünfzig Jahre Dieselmotor.*

CHAPTER NINE

Diesel, Eugen. "Aus der Kindheit des Dieselmotors," *Westermanns Monatshefte*, Vol. CXLIII (1928), 511–14.

———, and Georg Strössner. *Kampf um eine Maschine.*

Diesel, Rudolf. "Der Heutige Stand . . . ."

———. *Theorie und Konstruction eines rationellen Wärmemotors zum Ersatz der Dampfmaschinen und der heute bekannten Verbrennungs-motoren.* (Cited hereafter as *Theorie und Konstruction . . . .*)

"Diesel Cylinder Wear: A Symposium of Engineers," *The Power Engineer*, 1933.

Diesel Engines, Inc. *Fifty Years of Progress.*

Johanning, Albert N. P. *Der Diesel Motor, seine Entstehung und volkswirtschaftliche Bedeutung.* (Cited hereafter as *Der Diesel Motor . . . .*)

Maschinenfabrik Augsburg-Nürnberg. *25 Jahre Diesel-Kraftwagen, 1924–1949.*

———. *40 Jahre Dieselmotor.*

Ostwald, Walter. *Rudolf Diesel und die motorische Verbrennung.*

Pounder, C. C., ed. *Diesel Engine Principles and Practices.*

Serruys, Max. "Les moteurs et les turbines à gaz," *La revue général de mécanique.*

CHAPTERS TEN AND ELEVEN

Bennet, Merrill R., et al. *The First Fifty Years of the Diesel Engine in America.*

Büchi, Alfred J. *Early Days in Diesel Development.*

Diesel, Eugen. *Die Geschichte des Diesel-Personenwagens.*

Diesel, Rudolf, and Moritz Schröter. "Zwei Vorträge: Diesel's Rationeller Wärmemotor," *Zeitschrift des Verbandes Deutscher Ingenieure*, 1897.

Diesel Engine Users Association of London. *No. S–102*, "Coal, Synthetic Fuels and Oil from the National Standpoint."

Diesel Power and Equipment.

Meier, Edward Daniel. "Diesel's Rational Heat Motor," *Technology Quarterly*, 1899.

Meyer, Eugen. *Bericht über Versuche an Wärme-Motoren "Patent Diesel."*

Nickel, ———. *Diesel und seine Erfindung in der ausländischen Presse.*

Sulzer Brothers. *Historical Retrospect and Technical Development.*

CHAPTERS TWELVE AND THIRTEEN

Capitaine, Emil. *Kritik des Dieselmotors.*
Crew. "The Tragedy of Rudolf Diesel," *Scientific American,* December, 1940.
Diesel, Eugen. *Diesel.*
Diesel Engine Progress.
Diesel Engine Users Association of London. *No. S–84,* "Liquid Fuel from Coal."
——. *No. S–101,* "Injection, Ignition and Combustion in High Speed Heavy Oil Engines."
Ford, Louis R., ed. *Marine Diesel Engines.*
Goldbeck, Gustav. "Aus der Frühzeit des Dieselmotors," *Motortechnische Zeitschrift,* 1958.
Lüders, J. *Der Dieselmythus.*
Pounder. *Diesel Engine Principals and Practices.*
Rauck, M. J. B. *50 Jahre Dieselmotor.*
Rose, E. Mortimer. *Diesel Engine Design.*

CHAPTER FOURTEEN

Diesel, Eugen. *Völkerschicksal und Technik.*
Diesel, Rudolf. *Solidarismus.*
Diesel Engine Progress.
Emerson, Brian P., ed. *Diesel Power.*
Lehmann, Johannes. *M. S. Selandia.*
——. *Rudolf Diesel and Burmeister & Wain.*
Wadman, Rex W., ed. *Diesel and Gas Engine Progress in Industry, in Transportation, on the Sea, in the Air, Underground.*
Worsoe, Wilhelm. *Die Mitarbeit der Werke Fried. Krupp an der Entstehung des Dieselmotors.*

CHAPTER FIFTEEN

Camm, F. J., ed. *Diesel Vehicles.*
Codrington. *Shadows of Two Great Leaders.*
Diesel, Eugen. *Das Gefährliche Jahrhundert.*
——. *Wir und das Auto.*
Diesel, Rudolf. *Die Entstehung des Dieselmotors.*
Hallegan, Philip T. *Diesel Engine Material, Design and Performance.*
Hansfelder, Ludwig. *Die kompresserlose Dieselmaschine.*

Humm, Bruno. "Zur Erinnerung an Rudolf Diesel," *Storm und See,* 1958.

Powers. "Rudolf Diesel," *Marine Review,* Vol. LVIII (1928), 91–132.

Schildberger, Friedrich. "Der Weg der Erfindung Diesels bis zu unseren Tagen," *Bayerisches Braunkohlen-Bergbau,* 1958.

Siebertz, Paul. "Rudolf Diesel und der Automobilmotor," *Blätter für Technikgeschichte,* 1950.

CHAPTER SIXTEEN

Bennet, *et al. The First Fifty Years of the Diesel Engine in America.*

Diesel, Eugen, and Georg Strössner. *Kampf um eine Maschine.*

Diesel Engine Manufacturing Association. *Diesel Engine Standard Practices.*

The Diesel Motor.

"Dieselmotoren der Maschinenfabrik Augsburg-Nürnberg, A.G.," 1911.

Geiger, J., *et al. Diesel Maschinen.*

Grove's Dictionary of Music and Musicians. New York and London, The Macmillan Company.

L'Orange, Prosper. "Die Entwicklung des raschlaufenden Dieselmotors bis zum Kleinstmotor ohne Einspritzpumpe," *Motortechnische Zeitschrift,* 1939.

Nordberg Manufacturing Company. *A History of the Busch Sulzer Diesel Engine Company.*

Siebertz, Paul. *Karl Benz.*

CHAPTERS SEVENTEEN AND EIGHTEEN

Arkansas Sentinel, April 7, 1912.

Chalkley, A. P. *Diesel Engines for Land and Marine Work.*

Diesel, Eugen, and Georg Strössner. *Kampf um eine Maschine.*

Diesel, Rudolf. *The Present Status of the Diesel Engine . . . .*

Diesel Engine Manufacturers Association. *Standard Practices for Low and Medium Speed Stationary Diesel and Gas Engines.*

Diesel Engine Users Association of London. *No. S–86,* "Marine Oil Engines."

Documents and Records, Thomas Alva Edison Foundation, West Orange, New Jersey.

Gercke, Maximilian. *Stationary High Powered Internal Combustion Engines for Electric Generators.*

Gray, Paul Eastman. *The Diesel Electric Locomotive Dictionary.*

New York American, April 14, 1912.
Rosbloom, Julius. *Diesel Reference Guide.*
St. Louis *Post Dispatch,* April 4, 9–15, 1912.

CHAPTERS NINETEEN AND TWENTY

Associated Press coverage of the sinking of the S.S. *Titanic,* April, 1912.
Diesel, Eugen. *Die Macht des Vertrauens.*
———. *Ringen um Europa.*
Diesel, Rudolf. *Solidarismus.*
Documents and Records, Thomas Alva Edison Foundation.
Holmgren. "Rudolf Diesel, 1858–1913," *Marine Review,* Vol. LVIII (1928), 91–132.
London *Daily Mail,* October 1, 1913.
Manchester Guardian, September 30, 1913.
The New York Evening World, October 2, 1913.
The New York Times, October 2, 1913.
St. Louis *Post Dispatch,* April 12, 1912.
St. Louis *Star-Times,* April 13, 1912.

CHAPTERS TWENTY-ONE AND TWENTY-TWO

Adams, Orville. *Modern Diesel Engine Practice.*
Arbeitsgemeinschaft Dieselschienenverkehr, Göttingen. *Dieselfahrzeuge im Schienenverkehr.*
Bock, Siegfried. *Die Dieselmaschine in Land- und Schiffsbetrieb.*
Diesel, Eugen. *Die Geschichte des Diesel-Personenwagens.*
———. *Vom Verhängnis der Völker.*
Diesel, Rudolf. *Diesel's Rational Heat Engine.*
———. *The Present Status of the Diesel Engine . . . .*
———. *Theorie und Konstruction . . . .*
Diesel Engine Manufacturers Association. *Standards of the Diesel Engine Manufacturers Association.*
Diesel Engine Manufacturing Association. *Diesel Engine Standard Practices.*
Diesel Engineers Institute. *Diesel Engine Catalog.* Vols. I–XXIV.
Diesel Publication, Inc. *Diesel Engineer Handbook.*
"Dieselmotoren der Fried. Krupp, A. G. in Essen," 1925.
Gercke. *Stationary High Powered Internal Combustion Engines for Electric Generators.*
Johanning. *Der Diesel Motor . . . .*

Klöckner-Humboldt-Deutz, A.G. *Die Entwicklung des Dieselmotors in der Klöckner-Humboldt-Deutz, A.G.*

L'Orange, Prosper. *Ein Beitrag zur Entwicklung der kompresserlosen Dieselmotoren.*

Maxwell, Robert, ed. *Information U.S.S.R.*

Mayer, Max. *Diesel-Lokomotiven und Diesel-Triebwagen.*

Mayr, Fritz. *Ortsfeste Dieselmotoren und Schiffsdieselmotoren.*

Morrison, Lacey H. *American Diesel Engines.*

————. *Handbook on Diesel Engines.*

Porter, Geoffrey. *Notes on the Wear of Diesel Engines.*

Regenbogen, C. *Der Dieselmotorenbau auf der Germaniawerft.*

Richardson, Ralph A. *Diesel, the Modern Power.*

Schildberger, Friedrich. *Bosch und der Dieselmotor.*

Siebertz, Paul. *30 Jahre Mercedes-Benz-Diesel-Lastwagen.*

Wilkinson, Paul H. *Diesel Aviation Engines.*

Worthington Pump and Machinery Corporation. *Worthington and the First Fifty Years of American Diesel Engine.*

A Diesel Bibliography

Books and Pamphlets

Adams, Orville. *Modern Diesel Engine Practice: Theory, Practical Application, Operations, Maintenance, Repairs.* New York, Norman W. Henley, 1931.

Arbeitsgemeinschaft Dieselschienenverkehr, Göttingen. *Dieselfahrzeuge im Schienenverkehr. Eine vergleichende Betrachtung gegenüber Dampf- und elektrischer Zugförderung.* Darmstadt, Röhrig, 1954.

Baentsch, Erich. *Dieselmotoren-Praxis.* Berlin, Schiele & Schön, 1951.

Bennet, Merrill R., Ralph L. Boyer, C. G. A. Rosen, and C. E. Beck. *The First Fifty Years of the Diesel Engine in America.* New York, American Society of Mechanical Engineers, 1949.

Bock, Siegfried. *Die Dieselmaschine in Land- und Schiffsbetrieb.* Hamburg, Eckhardt & Messtorff, 1952.

Büchi, Alfred J. *Early Days in Diesel Development.* Railway Gazette, 1939. (At the Deutsches Museum, Munich.)

Camm, F. J., ed. *Diesel Vehicles: Operation, Maintenance and Repair.* London, G. Newnes, Ltd., 1940.

Capitaine, Emil. *Kritik des Dieselmotors.* Vortrag, Bezirks-Verein Deutscher Ingenieure zu Frankfurt a. M., am 20. April, 1898. Frankfurt, 1898. (At the Deutsches Museum, Munich.)

Chalkley, A. P. *Diesel Engines for Land and Marine Work.* With an introductory chapter by Rudolf Diesel. New York, Van Nostrand, 1912.

———. *Rudolf Diesel: A Detailed Biography and an Historical Survey of the Origin and Development of the Diesel Engine.* London, British Motor Ship, Vol. XVIII, 1937.

Clerk, Dugald. *The Gas and Oil Engine.* New York & Bombay, Longmans, Green, 1897.

Codrington, George W. *Shadows of Two Great Leaders—Rudolf Diesel and Alexander Winton.* New York, The Newcomen Society of England, American Branch, 1945.

Diesel, Eugen. *Diesel: Der Mensch, Das Werk, Das Schicksal.* Hamburg, Hanseatische Verlagsanstalt, 1937.

————. *Die erste Zündung. Wie der Dieselmotor entstand.* Hamburg, Hanseatische Verlagsanstalt, 1939.

————, Gustav Goldbeck, and Friedrich Schildberger. *From Engines to Autos: Five Pioneers in Engine Development and Their Contributions to the Automotive Industry.* Translated by Peter White. Chicago, Henry Regnery Company, 1960.

————. *Das Gefährliche Jahrhundert.* Berlin, Erich Schmidt, 1950.

————. *Germany and the Germans.* Translated by W. D. Robson-Scott. New York, The Macmillan Company, 1931.

————. *Die Geschichte des Diesel-Personenwagens.* Stuttgart, Deutsche Verlags-Anstalt, 1955.

————. *Jahrhundertwende. Gesehen im Schicksal meines Vaters.* Stuttgart, Reclam-Verlag, 1949.

———— and Georg Strössner. *Kampf um eine Maschine: Die Ersten Dieselmotoren in Amerika.* Berlin, Erich Schmidt, 1950.

————. *Die Macht des Vertrauens.* München, H. Rissen, 1946.

————. *Philosophie am Steuer.* Mit 46 Federzeichnungen von Willy Wedmann. Stuttgart, Reclam-Verlag, 1952.

————. *Ringen um Europa.* Leipzig, Bibliographisches Institute A. G., 1935.

————. *Schweizer Streiflichter.* Zürich, Rotapfel, 1953. (At the Deutsches Museum, Munich.)

————. *Die Söhne Fortunas; Tragödie in fünf Akten.* Stuttgart und Berlin, Cotta, 1925.

————. *Völkerschicksal und Technik.* Stuttgart und Berlin, Cotta, 1930.

————, Gustav Goldbeck, and Friedrich Schildberger. *Vom Motor zum Auto: Fünf Männer und ihr Werk.* Stuttgart, Deutsche Verlags-Anstalt, 1957.

————. *Vom Verhängnis der Völker: Das Gegenteil einer Utopie.* Stuttgart und Berlin, Cotta, 1934.

————. *Wir und das Auto: Der Motor verwandelt die Welt.* Mannheim, Bibliographisches Institut, 1933.

Diesel, Rudolf. *Antwort auf Emil Capitaine's Kritik des Dieselmotors.* München, 1898.

————. *Diesel's Rational Heat Engine: A Lecture.* New York, Progressive Age Publishing Company, 1897.

————. *Die Entstehung des Dieselmotors.* Berlin, J. Springer, 1913.

————. *The Present Status of the Diesel Engine in Europe and a Few Reminiscences of the Pioneer Work in America.* St. Louis, Busch-Sulzer Brothers Diesel Engine Company, 1912. A lecture delivered before the Associated Engineering Societies of St. Louis, April 13, 1912.

————. *Solidarismus: Natürliche Wirtschaftliche Erlösnung des Menschen.* Leipzig, R. Oldenbourg, 1903.

————. *Theorie und Konstruction eines rationellen Wärmemotors zum Ersatz der Dampfmaschinen und der heute bekannten Verbrennungs-motoren.* Berlin, J. Springer, 1893. (At the Deutsches Museum, Munich.)

————. *Theory and Construction of a Rational Heat Motor.* Translated by Bryan Donkin. London, E. & F. N. Spon; New York, Spon and Chamberlain, 1894.

Diesel Engine Manufacturers Association. *Standard Practices for Low and Medium Speed Stationary Diesel and Gas Engines.* Washington, 1958.

————. *Standards of the Diesel Engine Manufacturers Association.* Chicago, 1951.

Diesel Engine Manufacturing Association. *Diesel Engine Standard Practices.* New York, Diesel Publications, Inc., 1935.

————. *Marine Diesel Standard Practices.* Chicago, 1948.

Diesel Engine Progress. London, Tothill Press, 1955.

Diesel Engine Users Association of London. *Special Publications.*

No. S–83, "Report on Heavy-oil Engine Working Cost." 1926–27.

No. S–84, "Liquid Fuel from Coal." 1928.

No. S–85, "Some Considerations Regarding Peak-load Problems and High Powered Peak-load of Diesel Engines." 1928.

No. S–86, "Marine Oil Engines." 1928.

No. S–88, "Some Notes on High Speed Oil Engines." 1929.

No. S–89, "Repairs to Diesel Engine Parts by Electric-Deposition." 1929.

No. S–90, "Indicating Oil Engines." 1929.

No. S–91, "Modern Engineering Cast Irons and Properties." 1929.

No. S–92, "A Form of Coupling for Geared Diesel Engines for Land and Marine Purposes." 1929.

No. S–93, "High-Powered Oil Engines for Land Purposes." 1930.

No. S–94, "Report on Heavy-Oil Engine Working Costs." 1928–30.

No. S–95, "The High-Speed Diesel Engine." 1930.

No. S–96, "High Power Heavy Duty Oil Engines." 1930.

No. S–97, "Progress in the Application of the Diesel Engine to Road Transport." 1930.

No. S–98, "Waste Heat Recovery from Internal Combustion Engines with Particular Reference to Marine Oil Engines." 1930.

No. S–100, "The Development of the Multicylinder Horizontal Oil Engine." 1931.

No. S–101, "Injection, Ignition, and Combustion in High Speed Heavy Oil Engines." 1931.

No. S–102, "Coal, Synthetic Fuels and Oil from the National Standpoint." 1931.

Diesel Engineers Institute. *Diesel Engine Catalog,* Vol. I–XXIV (1936–59). Los Angeles, Diesel Publishing Company.

Diesel Engines, Inc. *Fifty Years of Progress.* New York, 1948.

The Diesel Motor: Diesel Motor Company of America. New York, Bartlett & Company, 1898.

Diesel Power and Equipment. Vancouver, British Columbia, Westrack Publications, Ltd., 1960.

Diesel Publications, Inc. *Data Service: Diesel Power and Direct Transportation.* New York, 1939.

———. *Diesel Engineer Handbook.* New York, 1943.

Emerson, Brian, ed. *Diesel Power: Fuel Ignition Systems.* New York, Diesel Publications, Inc., 1956.

Ensslein, Max. *Explosionsmotor und Verbrennungsmotor (Ottomotor und Dieselmotor).* Stuttgart, Vereins Buchdruckerei, 1900.

Ford, Louis R. *Questions and Answers for Marine Diesel Engines.* New York, Diesel Publications, Inc., 1942.

Ford, Louis R., ed. *Marine Diesel Engines.* New York, Diesel Publications, Inc., 1943.

Gedenkheft zum 100. Geburtstag von Rudolf Diesel. Stuttgart, Motortechnische Zeitschrift, 1958.

Geiger, J. H. Hintz, and E. Hubendeck. *Diesel Maschinen*. Berlin, 1926.

Gercke, Maximilian. *Stationary High Powered Internal Combustion Engines for Electric Generators*. London, 1930.

Gray, Paul Eastman. *The Diesel Electric Locomotive Dictionary*. Lyons, Ill., Privately published, 1938.

Gumilevskij, Leo. *Rudolf Diesel. Ego zizn' i dejatel'nost' biografičeskij očerk*. (His Life and Work) Moskva, Leningrad, Gosudarstvennoe exergeticeskoe izdatel' stvo, 1934.

Hallegan, Philip T. *Diesel Engine Material, Design and Performance*. London, Diesel Engine Users Association, 1947.

Hansfelder, Ludwig. *Die kompresserlose Dieselmaschine. Ihre Entwicklung auf Grund der In- und Ausländischen Patent-Literatur*. Berlin, M. Krayn, 1928.

Johanning, Albert N. P. *Der Diesel Motor, seine Entstehung und volkswirtschaftliche Bedeutung*. Nürnberg, Wilhelm Tümmel, 1901. (At the Deutsches Museum, Munich.)

Klöckner-Humboldt-Deutz, A. G. *Die Entwicklung des Dieselmotors in der Klöckner-Humboldt-Deutz, A. G.* Köln, 1954.

Kraus, Wolfgang. *Rudolf Diesel*. Nürnberg, Olympia Verlag, 1949.

Kurzel-Runtscheiner, Erich von. *Der Dieselmotor. Eine Deutsche Erfindung erobert die Welt*. Fromme Studentenkalender, 1942–43.

Lehmann, Johannes. *M.S. Selandia*. Copenhagen, 1937.

———. *Rudolf Diesel and Burmeister & Wain*. Copenhagen, 1938.

L'Orange, Prosper. *Ein Beitrag zur Entwicklung der kompresserlosen Dieselmotoren*. Berlin, Richard Carl Schmidt & Company, 1934.

Lossow, P. von, E. Meyer, M. Schröter, Rudolf Diesel, and C. Boccali. *Neuere Literatur über den Diesel-Motor*. Berlin, A. W. Schade, 1903. (At the Deutsches Museum, Munich.)

Lüders, J. *Der Dieselmythus. Quellenmässige Geschichte der Entstehung des heutigen Ölmotors*. Berlin, M. Krayn, 1913.

Maschinenfabrik Augsburg-Nürnberg. *Bilder, Dokumente und Erläuterungen zur Entstehung des Diesel-Motors unter besonderer Berücksichtigung der Entwicklung des Fahrzeug-Diesel-Motors*. (Zum 100. Geburtstag von Rudolf Diesel, 18. März 1958.) München, 1958.

———. *Denkschrift zur Erinnerung an die Feier der Stadt Augsburg am Anlass der Enthüllung einer Rudolf Diesel Gedenktafel und an die Feierstunde der Maschinenfabrik Augsburg-Nürnberg A.G. anlässig der Eröffnung des M.A.N. Werkmuseums in Augsburg am 23. April 1953.* Augsburg, Selbstverlag, 1953.

——. *25 Jahre Diesel-Kraftwagen, 1924–1949.* Nürnberg, Selbstverlag, 1949.

——. *Fünfzig Jahre Dieselmotor. 1897–1947.* 1948.

——. *40 Jahre Dieselmotor.* Augsburg, 1937.

Maxwell, Robert, ed. *Information U.S.S.R.,* Vol. I. *Countries of the World, Information Series.* New York, The Macmillan Company, 1962.

Mayer, Max. *Diesel-Lokomotiven und Diesel-Triebwagen. Denkschrift zum 100-jährigen Bestehen der Maschinenfabrik Esslingen.* Esslingen, Selbstverlag, 1949.

Mayr, Fritz. *Ortsfeste Dieselmotoren und Schiffsdieselmotoren.* Vienna, J. Springer, 1960.

Meyer, Eugen. *Bericht über Versuche an Wärme-Motoren "Patent Diesel."* Charlottenburg, 1900. (At the Deutsches Museum, Munich.)

Meyer, Paul. *Beiträge zur Geschichte des Dieselmotors.* Berlin, J. Springer, 1913.

Morrison, Lacey H. *American Diesel Engines.* New York, McGraw-Hill Book Company, 1939.

——. *Handbook on Diesel Engines.* New York, Diesel Publications, Inc., 1938.

——. *Diesel Engine Handbook,* Vol. XI. New York, Diesel Publications, Inc., 1941.

——. *Diesel Engineer, A Complete Home Study Course.* Revised by Charles F. Foell. New York, Diesel Publications, Inc., 1941.

Neumann, Kurt. *Untersuchungen an der Dieselmaschine.* (Forschungsarbeiten auf dem Gebiete des Ingenieurswesens. H. 245). Berlin, 1921.

Newell, A. B. *Vest Pocket Diesel Manual.* New York, Diesel Publications, Inc., 1946.

Nickel,——. *Diesel und seine Erfindung in der ausländischen Presse.* 1939.

Nitske, W. Robert. *The Complete Mercedes Story.* New York, The Macmillan Company, 1956.

Nordberg Manufacturing Company. *A History of the Busch Sulzer Diesel Engine Company.* St. Louis.

Ostwald, Walter. *Rudolf Diesel und die motorische Verbrennung.* München, R. Oldenbourg, 1956.

Pachtner, Fritz. *Patent 67207. Rudolf Diesel und das Werk seines Lebens.* Berlin, W. Limpert, 1943.

Porter, Geoffrey. *Notes on the Wear of Diesel Engines.* London, Diesel Engines Users Association, 1920.

Pounder, C. C., ed. *Diesel Engine Principles and Practices.* New York, New York Philosophical Library, 1955.

Rauck, Max J. B. *50 Jahre Dieselmotor.* München, Leibniz Verlag, 1949.

Regenbogen, C. *Der Dieselmotorenbau auf der Germaniawerft.* Essen, Fried. Krupp Jahrbuch, 1913.

Richardson, Ralph A. *Diesel, the Modern Power.* Detroit, General Motors Corporation, 1950.

Rosbloom, Julius. *Diesel Reference Guide.* Jersey City, New Jersey Industrial Institute, 1931.

Rose, E. Mortimer. *Diesel Engine Design.* Manchester and London, Emmott & Company, 1917.

Sass, Friedrich. *Geschichte des deutschen Verbrennungsmotorenbaues von 1860 bis 1918.* Berlin, Göttingen, Heidelberg, Springer-Verlag OHG, 1962.

Schildberger, Friedrich. *Bosch und der Dieselmotor.* Stuttgart, Stähle & Friedel, 1950.

Siebertz, Paul. *30 Jahre Mercedes-Benz-Diesel-Lastwagen. Zur Geschichte des Fahrzeug Dieselmotors.* Stuttgart, Daimler-Benz A. G., 1953.

———. *Karl Benz: Ein Pioneer der Motorisierung.* Stuttgart, Reclam Verlag, 1950.

Sulzer Brothers. *Historical Retrospect and Technical Developments.* Winterthur, Switzerland.

Tanzer, Karl. *Rudolf Diesel's Motor und Schicksal.* Wien-Mödling, St. Gabriel Verlag, 1958.

Wadman, Rex W., ed. *Diesel and Gas Engine Progress in Industry, in Transportation, on the Sea, in the Air, Underground.* New York, Diesel Engines, Inc., 1949.

Wilkinson, Paul H. *Diesel Aviation Engines.* New York, National Aeronautics Council, 1942.

Worsoe, Wilhelm. *Die Mitarbeit der Werke Fried. Krupp an der Entstehung des Dieselmotors.* Kiel, 1940.

Worthington Pump and Machinery Corporation. *Worthington and the First Fifty Years of American Diesel Engine.* Harrison, New Jersey.

Articles and Periodicals

Crew, Henry. "The Tragedy of Rudolf Diesel," *Scientific American,* December, 1940.

Diesel, Eugen. "Aus der Kindheit des Dieselmotors," *Westermanns Monatshefte* (Braunschweig), Vol. CXLIII (1928), 511–14.

Diesel, Rudolf. "The Diesel Engine," *Progressive Age,* Vol. XVII (1899).

————. "Der Heutige Stand der Wärmekraftmaschinen und die Frage der Flüssigen Brennstoffe, unter besonderer Berücksichtigung des Diesel-Motors," *Zeitschrift des Vereins Deutscher Ingenieure* (Berlin), Vol. XLVII (1903), 1366–75.

————. "Der Mechanische Wirkungsgrad und die undizierte Leistung der Gasmaschine," *Zeitschrift des Vereins Deutscher Ingenieure* (Berlin), Vol. XLIX (1905), 449.

————, and Moritz Schröter. "Zwei Vorträge: Diesel's Rationeller Wärmemotor," *Zeitschrift des Verbandes Deutscher Ingenieure* (1897). (At the Deutsches Museum, Munich.)

"Diesel Cylinder Wear: A Symposium of Engineers," *The Power Engineer* (London), 1933.

Diesel Power, June, 1925–June, 1961. A periodical published at Stamford, Connecticut.

"Dieselmotoren der Fried. Krupp, A.G. in Essen" (1925). (At the Deutsches Museum, Munich.)

"Dieselmotoren der Maschinenfabrik Augsburg-Nürnberg, A.G." (1911). (At the Deutsches Museum, Munich.)

Fürst, Arthur. "Ein Erfinderschicksal," *Uhu Zeitschrift,* 1925. (At the Deutsches Museum, Munich.)

Goldbeck, Gustav. "Aus der Frühzeit des Dieselmotors," *Motortechnische Zeitschrift* (Stuttgart), 1958. (At the Deutsches Museum, Munich.)

————. "Rudolf Diesel zum 100. Geburtstag," *VDI-Nachrichten,* 1958.

Holmgren, E. J. "Rudolf Diesel, 1858–1913," *Nature,* 1958. (At the Deutsches Museum, Munich.)

Humm, Bruno. "Zur Erinnerung an Rudolf Diesel," *Storm und See,* 1958. (At the Deutsches Museum, Munich.)

"Ingenieur Diesel, Erfinder des Dieselmotors in München," 1900. (At the Deutsches Museum, Munich.)

Kennedy, John B. "The Cavalcade of Diesel," *Universal Engineer* (New York), January, 1937, pp. 23–47.

L'Orange, Prosper. "Die Entwicklung des raschlaufenden Dieselmotors bis zum Kleinstmotor ohne Einspritzpumpe," *Motortechnische Zeitschrift* (Stuttgart), 1939.

Martinaglia, L. "Aus der Entwicklungszeit des Dieselmotors," *Schweizerische Bauzeitung,* 1948.

Meier, Edward Daniel. "Diesel's Rational Heat Motor," *Technology Quarterly* (Boston), 1899.

"Mitteilungen über den Dieselmotor," Vortrag Abdruck in der *Zeitschrift des Oesterreichischen Ingenieur-und Architekt Vereins,* 1901. (At the Deutsches Museum, Munich.)

"Les Moteurs Diesel, deurs applications industrielles," *La Technique Moderne,* 1954.

"Die Motorschiffahrt in den Kolonien," (Berlin), 1911. (At the Deutsches Museum, Munich.)

Peyonnet, J. "Diesel et la conquête de l'énergie," *Musée du Conservatoire des Art et Métiers* (Paris). Mars–Avril, 1959.

Powers, E. C. "Rudolf Diesel," *Marine Review* (Cleveland), Vol. LVIII (1928), 91–132.

Reichelt, Johannes. "Rudolf Diesel. Ein Beitrag zum Gedenken anlässig des 100. Geburtstages," *Schiffbautechnik,* 1958.

Richter, Ludwig. "Rudolf Diesel und seine Bedeutung," *VDF Zeitschrift,* 1958. (At the Deutsches Museum, Munich.)

Schildberger, Friedrich. "Der Weg der Erfindung Diesels bis zu unseren Tagen," *Bayerisches Braunkohlen-Bergbau,* 1958. (At the Deutsches Museum, Munich.)

Schnauffer, Kurt. "Die Erfindung Rudolf Diesels — Triumph einer Theorie," *VDF Zeitschrift,* 1958. (At the Deutsches Musuem, Munich.)

———. ". . . die Wärme unser Brennstoffe rationeller auszunützen,"

Motortechnische Zeitschrift (Stuttgart), 1958. (At the Deutsches Museum, Munich.)

Serruys, Max. "Les moteurs et les turbines à gaz," *La revue général de mécanique,* Septembre–Octobre, 1950.

Siebertz, Paul. "Rudolf Diesel und der Automobilmotor," *Blätter für Technikgeschichte,* 1950. (At the Deutsches Museum, Munich.)

Sittauer, Hans Leo. "Diesel: Eine Erfindung erobert sich die Welt," *VEB Verlag für Verkehrswesen* (Berlin), 1961.

Stead, H. V. "Forty Years Onward: The Fortieth Anniversary of Dr. Diesel's First Heavy Oil Engine," *Gas and Oil Power,* Vol. XXXII (1937).

Sulzer. "50 Jahre Sulzer-Dieselmotoren," *Technische Rundschau* (1947).

"Viertakt-Kohlestaub-Kraftmaschine," *ADAC Motorwelt,* 1927. (At the Deutsches Museum, Munich.)

Index

Adagon (gunboat) : 213

Aktieselskabet Burmeister and Wain's *Maskin- og Skibsbyggeri:* 126, 168–71, 276–77; license from Diesel acquired by, 169

Allgemeine Gesellschaft für Dieselmotoren, A.G.: 134, 136, 149, 195

Allis-Chalmers: 255

American Buda Company: 258

American Diesel Engine Company: 150–51, 251–52, 253

American Locomotive Company: 256, 259

American Society of Mechanical Engineers: 178, 208; Diesel made honorary member of, 205; lecture of Diesel to, 231

Ammonia engine: 76, 77

Amundsen, Roald: 168, 202

Anheuser-Busch brewery: 226, 252

Aorangi (liner) : 271

Arkansas Sentinel: 216

Atchison, Topeka and Santa Fe Railroad: 260

Atlas Imperial Diesel Engine Company: 256, 258

Augsburg, Bavaria: 5, 8, 9, 11–14, 38, 92, 104, 114, 120, 269

Augsburg-Krupp consortium: 96, 100–101, 106, 108, 115, 116, 117, 119, 120, 127, 144, 182, 185; *see also Maschinenfabrik* Augsburg and Fried. Krupp *Werke*

Automobil-Aussetellung (Berlin, 1924) : 264

Baden-Baden, Baden: 125

Baeyer, Adolf von: 119

Baku (oil fields) : 138, 173–74

Baldwin Locomotive Works: 251

Bar-le-Duc, France: 97, 122, 166

Barnickel, Betty (cousin) : 35, 37, 44, 51

Barnickel, Christoph (cousin and brother-in-law) : 35, 37, 44; courts Emma Diesel, 62; builds home, 92

Barnickel, Emma Diesel (sister): builds home, 92; last visit of brother to, 237

Baumgarten and Burmeister: 168; *see also Aktieselskabet* Burmeister and Wain

Bavaria: 9

Bayerische Hypotheken und Wechselbank: 237

Bayerische Motoren Werke: 279
Beardmore, William, and Company: 281
Bennett, Gordon: 189
Benz & Cie.: 188, 264, 270, 278, 279
Benz automobile: 70
Benz, Bertha: 201
Benz, Karl: 83, 132, 155, 201
Berengaria (liner) : 234
Berlin, Germany: 223, 236; Diesel moves to, 73–74, 92
Bernardi, Lauro: 276
Berndorfer *Metallwarenfabrik:* 182
Bismarck-Schönhausen, Prince Otto von: 75, 92
Blohm and Voss Company: 280
Bodensee (Lake Constance): 155, 163
Bodet, Adrien: 212
Bosch, Robert: 83, 99, 189
Böttcher, Anton: supervises installations, 121, 129, 136; installs first Diesel engine in U.S., 179–80
Bristol Aeroplane Company: 281
British Diesel Company: 253
British Diesel Engine Company: *see* Diesel Engine Company of England
Brons patents: 258
Brown, Erich: 121
Büchi method: 257, 268, 273
Budd rail car: 217
Burlington Railroad: 255, 260
Burmeister and Wain: *see Aktieselskabet* Burmeister and Wain
Busch, Adolphus: 124–26, 201; Meier sent to Augsburg by, 125; goes to Germany, 125–26; company difficulties of, 150–51;

engine development by, 177–80; contract with Diesel, 179; Diesel's visit to, 192–95, 203–205, 209, 211; employs press agent, 205, 226; entertains Diesels, 229; closes New York office, 252–53; founds new company, 253–54
Busch–Sulzer Brothers Diesel Engine Company: 202, 226, 249, 254, 256–57
Buz, Heinrich: 57, 105; Diesel contact with, 77; comment of on first Diesel Engine, 90; attends directors meeting, 100, 108; agrees with Diesel, 115–16; meets Meier, 125; Busch confides to, 253

C. and C. Electric Company: 180
Canada: 261, 274
Capitaine, Emil: 77, 115; patent controversy, 129–30
Carels Frères: 97, 167
Carels, George: letter of to Jean, 238; invites Diesel, 240; leaves for England with Diesel, 241, 244–46; reassures Diesel, 244; reports disappearance of Diesel, 245
Carels, Jean: 238
Carlsund, Anton: 127
Carnot, Sadi: 50
Caspian Sea: 174, 213
Caterpillar Tractor Company: 258–59
Christian X (formerly *Fionia,* liner) : 171, 202
Churchill, Winston: 170
Civil War (U.S.) : 152, 178; German divisions, 21
Clerget, Pierre: 280
Clerk, Douglas: 33

Cody, William (Buffalo Bill): 178

Coertsen (pilot steamer), reports floating body: 246

Cohen, Louis Phillipe, partnership with Diesel: 64

Collier Trophy: 283

Cologne, Germany: 36, 279

Columbian Exposition (Chicago): 47

Conservatoire des Arts et Métiers (Paris): 18, 62, 198

Continental Motors Corporation: 259

Cooper Union (New York City), Diesel lectures at: 208, 228

Copenhagen, Denmark: 168-70

Cornelius (clipper ship): 167

Cornell University, Diesel lectures at: 208, 228

Cotta, Freiherr von (engineer): 39, 40

Courtens, Frans: 159, 244

Cowles, V. B., and Company: 252

Cugnot, Nicolas Joseph, builds first self-propelled vehicle: 18

Cummins Engine Company: 258

Cunard Steamship Line: 31

Curie, Marie: 152

Curie, Pierre: 152

Czischek, Ludwig, seeks Diesel's advice: 186

Daimler, Gottlieb: 83, 115, 132, 155, 184

Daimler-Benz A.G.:270-71

Daimler engines: 70, 76

Daimler Motorengesellschaft: 270, 279

Danneborg (yacht): 169

De La Vergne Refrigerating Company: 251

Deschamp, Heinrich: 187

Detroit Air Show (1929): 282

Deutsche Lufthansa: 279

Deutschland (submarine): 267

Deutz *Gasmotorenfabrik: see Gasmotorenfabrik* Deutz

Diesel, Elise Strobel (mother): 13, 14; describes London and Channel to father, 15; in Paris meets T. Diesel, 15; marriage of, 15; birth of children, 16; as business partner, 19; leaves Paris for London, 25-26; sends son to relatives, 35; returns with family to Paris, 41; interest in supernatural, 44; Martha visits, 62; son gives spending money to, 114; lives alone, 198

Diesel, Emma (sister): birth of, 16; letter to, 46; marriage of, 62; *see also* Barnickel, Emma Diesel

Diesel, Eugen (son): 248; birth of, 69; childhood of, 91; goes to exhibition, 130-32; in new home, 158-61; recalls father's piano playing, 159; recounts vacations, 197; grows closer to father, 198; tells of visit to Edison, 231-34; "studies fund" of, 237; follows father's footsteps, 239; personal effects found on body identified by, 246; accepts father's death as suicide, 247

Diesel, Gottlieb Theodor Hermann: *see* Diesel, Theodor (father)

Diesel, Hanns Christoff (ancestor), first to use name of Diesel: 10

Diesel, Hedy (daughter): birth of, 65; childhood of, 91; goes to exhibition, 130–32; trip with father, 199; marriage of, 198–99, 237; birth of first daughter to, 235; visit with father, 241–42

Diesel, Johann Christoph (Johann I) (ancestor), first bookbinder and publisher in Memmingen: 10

Diesel, Johann Christoph (Johann II) (ancestor): president of bookmakers' guild, 10; death of, 10

Diesel Johann Christoph (Johann III) (paternal grandfather): birth of, 10; journeys by foot, 11; marriage of to Sabine Riess, 11; birth of two sons, 11; collector of butterflies, 12; death of, 12; withdrawal from realities, 239

Diesel, Louise (sister): birth of, 16; teaches music, 25, 35; death of, 44

Diesel, Martha Flasche (wife): background of, 60–61; meets Rudolf Diesel, 60; marriage of, 64; birth of first son, 64; birth of daughter, 65; birth of second son, 69; social standing, 76; suggests publication, 78; shares husband with *Schwarze Geliebte*, 83; letter to husband, 86; with family in Berlin, 91; sees first engine operating, 96; moves to Munich, 110; shopping and opera, 111; social life, 113–14; goes to exhibition with family, 130–32; social advantages, 145; deplores absences of husband, 145; plans magnificent home, 146; letter to husband, 148; builds home,

157–62; distant relationship with husband, 160; reaction to husband's construction office in home, 187; family trip to France, 197; trip to Russia, 200; resumes interest in husband's work, 201; accompanies husband to U.S., 203–207, 227–34; reacts to reporters, 205; social events in St. Louis, 229; accompanies husband to visit Edison, 231–34; birth of granddaughter, 235; travels with husband to Sicily, 236; visits mother, 241; visits daughter, 241; last gift from husband, 242, 246; last letter from husband, 243–44; accepts husband's death as suicide, 247

Diesel, Nikolaus Augustin (ancestor), messenger: 10

Diesel, Rudolf (son): 248; birth of, 64; childhood of, 91; goes to exhibition, 130–32; in new home, 158; remains aloof, 198; becomes recluse, 239; visit with father, 241; letter from father to, 244; accepts father's death as suicide, 247

Diesel, Rudolf (uncle): 11; in Paris, 13; returns to Augsburg, 13; gives advice, 35, 36

Diesel, Rudolf Christian Karl: in Paris, 6, 7; birth certificate of, 6; forebears of, 9; birth of, 16; boyhood of, 16, 23, 26–28, 31, 33; attends school, 17; medal awarded to, 17; visits museums, 18; learns languages, 22; reaction of to English factories, 34; sent by parents to Augsburg, 35–36; reception of by Barnickels, 36–37; studies in Augsburg, 38; inclined toward art, 40; attends trade school, 41; confirmation of, 42; decision to become engineer, 42; examination at trade school, 43; reunion of with family, 43–44; resolves to learn piano, 44; attends Polytechnic High School, 46; letter of to sister, 46; bronze medal awarded to, 47; receives German citizenship, 47; origin of idea of Diesel engine, 48, 49; health and habits, 49; enjoys music and opera, 51, 111, 113, 197; has "typhus," 51; achieves highest grade ever, 52; first factory job, 52, 271; first work assignment, 53; foresees use of small engine, 54; to Paris, 55; first patent of, 56; rumor of mistress, 56–57; *ingénieur civil*, 58; buys Stetson hat, 59; meets Martha Flasche, 60; letters to Martha, 62–64, 102–103, 123, 148; marriage of, 64; birth of son, 64; business failure of, 64; bottles gas for military weapon, 65; birth of daughter, 65; ammonia engine, 66; illness (headaches and gout, nervous exhaustion) 66, 84, 119, 123, 128–29, 146, 162–63, 196, 200, 230; outing in Allgäu, 67; idea of solar engine, 68; birth of second son, 69; before International Congress, 70; leaves Paris for Germany, 71; ice making, 72; moves to Berlin, 73; neighbors and friends, 74; finances, 74; reaction to Marx, 75; first patent, 77; seeks underwriting, 77; controversy

over publication, 78; second patent, 81; division of affection, 83; letters, 84; signs agreements for sales of engine, 85; arrives in Augsburg, 86–87; first engine tested, 87–90; with family in Berlin, 90–91; guest of the Barnickels, 92; second engine, 93–96; silent salute to, 96; brings Martha to see engine achievement, 96; travels to France and Belgium, 96–97; problems with engine, 97–100; visits Bosch, 99; admits lack of success, 100; seeks reassurance from wife, 100; experiments with illuminating gas, 100; demonstrates engine in Austria, 102; revises engine, 104–107; receives patents, 105, 110, 142; tests Lauster, 109; moves to Munich, 110–11; Christmas shopping, 111; letter of about Munich, 113; relatives supported by, 113, 114; ignores accusations, 115; meets Köhler, 116; acquires telephone and typewriter, 118–19; disagrees with Vogel, 118; dislike of Krumper by, 120; goes to Britain, 120–21, 128–29; goes to Switzerland, 121–22; receives Bavarian Michel medal, 122; lectures of, 122, 196; letter to Noé, 123; visits Busch, 126–28; sells rights for engines to U.S., 126; foreign representatives visit, 126; grants Russian rights to Nobel, 128; constant travel to ease problems, 129; refuses demand of Capitaine and enmity ensues, 130; enjoys exhibition with family,

130–32; founds new company, 134; becomes millionaire, 135; experiments with crude oil, 136–40; wins approval for more tests, 140; permits Pawlikowski to form own company, 141; income from consortium, 144; treatment of workers, 145; family grows away from, 145; plans magnificent home, 146; rest cure at Meran, 146–48; financial losses, 149; signs away patents, 149; bolstered flagging enthusiasm, 151; in relation to other inventors, 152–55; first successful dirigible, 155; builds home, 157–62; likes painting, 159, 244; plays piano for family, 159; seldom sees wife, 160–61; dislikes new home, 162, 240; rest cure on Bodensee, 163–65; *Solidarismus,* 164–65, 197; problems of bankruptcy threat, 166–71; rides watercraft powered by Diesel engine, 167; license to Burmeister and Wain, 169; wears special shoe, 172; receives U.S. patent, 178; contract with Busch, 179; sales franchises (by 1901), 185; confidence in engine for seacraft, 186; moved construction office to home, 186–88; auto racing fan, 189; buys automobile, 190; visits U.S., 190–94; impressions of U.S., 192–94; financial troubles, 194–96; litigation against, 196; takes vacations in Italy and France with family, 197; trip with Hedy, 199; begins "Origin" book, 200; financial deals, 200; trip with wife to Russia, 200; trip with wife to U.S., 203–207, 226–34; foresees air pollution, 206; captures headlines, 206; starts work on locomotive engine, 208; lecture appearances, 208–11; visits Arkansas, 216–25; description of, 218–19; enjoys soda pop, 224; newspaper interviews, 227–29, 230–31; private opinions of, 227; states four American virtues, 228; guest of honor, absent, 229–30; lectures at Cornell, 230; lectures at Annapolis, 230; lectures at Society of Engineers at Cooper Union, 231; meets Edison, 231–34; incompatability to Edison, 232; opinion of Edison, 234; birth of granddaughter, 235; real estate litigation, 235; mortgages home, 235; spurns counsel of bankers, 236; publication of history of engine, 236; travels to Sicily, 236; reaction of Frau Sulzer to, 237; temper flares up, 237; trip in dirigible, 237; consultant to British factory, 238; hosts receptions for engineers, 238; depression of, 239; dedicates book to Martha, 240; drives recklessly, 240, 242; strange behavior of, 241, 247; visits daughter, 241; parting gift to wife, 242; last reading of, 242; arrives in Ghent, 243; last correspondence, 243, 244; boards Channel ferry, 244; fails to meet friends for breakfast, 245; condition of personal effects in cabin of ferry, 245–46; floating body identified,

246; money and property at time of death, 246; family accepts death of as suicide, 247; no will of, 248; rumors concerning death of, 249

Diesel, Sabine Riess (paternal grandmother): 11

Diesel, Theodor (father): in Paris, 5–9, 13, 16; birth of, 9; in Augsburg, 12, 13; marriage of, 15; invents "light shedder," 19; leaves Paris, 24; with family in London, 25; in England, 35; returns to Paris, 41; interest in metaphysics, 44; *Heilmagnetiseur*, 47; Martha Flasche visits with, 62; quits practicing, 114; suggests treatments, 128; death of, 198; withdraws from reality, 239

Diesel engine: first patent, 77; first engine tested, 87–90; comments on tests, 90; second engine tests, 93–96; problems of engine, 97; experiments with illuminating gas, 100; demonstration in Austria, 102; revised model, 104–107; friction brake, 106; first commercial engine designed, 108–109; no accident record, 110; new engine completed, 111–12; first Swiss built, 122; spreads to U.S., 124; manufacturing rights for U.S., 126; first in ship, 127; first commercial installation, 127; first public display, 132–34; A-Motor, B-Motor, 134; internationalization of, 135; experiments with crude oil, 136–42; patent for compound engine, 142; failure of two-cycle engine,

150; equips Hindenburg 152; *grand prix,* 166; first in Russia, 173; in Russia, 172–73, 181, 201; in France, 174; first submarine engine, 175; in Britain, 175–76; in Switzerland, 175–76; in United States, 177–80, 227; in Canada, 177; advertising text, 180; in Germany, 181; sales franchises (by 1901), 185; "petite model" at fair, 187; first for automobile, 187–89; first for trucks, 188; first automobile production with, 189; total horsepower in use, 206; for locomotives, 208; development of in U.S., 250–62; in Great Britain, 253, 274–75; in Canada, 261; in U.S.S.R., 262, 263, 265, 266, 267; in Germany, 263–71; in Switzerland, 271–73; in the Netherlands, 273–74; in Italy, 275–76; in Denmark, 276–77; aero engine, development in Germany, 278–80; in France, 280–81, 282; altitude records, 281; in Great Britain, 281–82; in U.S. 282–83; in Czechoslovakia, 283; in Italy, 283

Diesel Engine Company of England: 167, 238, 240, 248

Diesel Engine Company of London: 176

Diesel Engine Works (Bar-le-Duc): 138

Diesel family: origin of and evolution of name, 9, 10

Diesel Motor Company of America: 179, 185, 192, 195, 202, 204, 205

Dieselmotorenfabrik Augsburg: 127, 149–50; failure of, 195

Diesels-Motorer of Stockholm: 147

Dietrich, Karl: 129, 136

Dohner, Herman: 282

Döpp, Professor von: 128

Dornier Company: 280

Drake, Freddy: 221, 222

Dreadnaught (battleship): 175

Dresden (ferry): boarded by Diesel, 244; held for search, 245

Dyckhoff, Frédéric (friend and engineer): meets Diesel, 97; constructed engine, 100, 151, 212; visits Diesel, 114, 122; tests oil, 138; called to appraise design, 252

East Asiatic Company, orders first major cargo vessel: 169

East India Company: 171

Ebbs, Hermann: 108

Edison, Mrs. Thomas Alva: 232, 233

Edison, Thomas Alva: 32, 155, 206; meets Diesel, 231–34; views of on European visit, 232; on engines, 233; opinion of Diesel, 234

Electrical exposition (New York, 1898): 180

Electro-technic Exposition (Frankfurt): 76

Ellis Island (New York): 191

Ellis, John: 224

Emmerich, Rhineland: 36

English Channel: 14, 24–25, 243, 280

Ensslein, Max: 136

Ericsson, John: 154

Erie Railroad: 261

Erney, Hans: 129

Esslingen *Werke:* 271

Fairbanks, Morse and Company: 255

Fairfield Shipbuilding and Engineering Company: 271

Fayetteville, Arkansas: 216, 218, 271

Fedden, Roy: 281

Fiat (Fabrica Italiana Automobili Torino): 276, 283

Fionia (cargo vessel): 169, 170

Flasche, Martha: as governess in Paris, 58; family of, 60; education of, 60, 61; courted by Diesel, 60–61; visits Diesel family, 62; marriage of, 64; *see also* Diesel, Martha Flasche

Ford, Henry: 255; overture to Diesel, 237–38

Fram (Arctic ship): 168

France: 167, 174, 207, 213, 225, 280

Franco-Prussian War: 23, 41

Frankfurt, Germany: 36, 76, 241, 243

Franz, Adolf: 223–24

Frichs, A/S, of Aarhus: 277

Fried. Krupp *Werke:* 15, 127, 178, 181, 214, 271; first forge, 28; decision to construct Diesel engine, 82; signs sales contract, 85; comment on engine, 90, 96; directors meeting, 100–101; conference, 108; signs contract with Deutz, 116–17; franchise, 120; exhibits engine, 132–34; gets early engine, 134; amount spent by, 144; *Grusonwerke,* 181–82; *Germaniawerft,* 183; role of in developing Diesel engine, 266–68

Friedrichshafen, Germany: 155, 163

Frisco Railroad: 216–17, 221

Frith, Arthur: 179
Fulda, Germany: 27
Fulton, Robert: 30, 31
Fulton Iron Works: 254

Galicia (oil fields): 138, 173, 181
Gasmotorenfabrik Deutz: 78, 114–15, 254, 268; questions Diesel patents, 115; signs contract with consortium, 116; exhibits engine, 132–34; sends engine to U.S., 179; engine development, 184–85; becomes Klöckner-Humboldt-Deutz A.G., 269
Gauss, Karl Friedrich: 154
Gebrüder Sulzer *Maschinenfabrik:* 52, 53, 56, 57, 70, 167, 198, 201, 214, 241, 253–54, 257, 271; decides to construct Diesel engine, 82; signs contract for sales of engine, 85; Diesel visit to, 121–22; exhibits engine, 132–34; engine development of, 175–76; marine engines, 202; builds large engines, 208; present factory locations of, 272
General Electric Company: 259
General Motors Corporation: 255
General Society for Diesel Engines: *see Allgemeine Gesellschaft für Dieselmotoren,* A.G.
German Navy: 266, 268
German-American Petroleum Corporation: 267
Germaniawerft: 183, 191, 266
Germany: 138, 163, 203, 207, 225, 280; Diesel engines in, 263–71
Ghent, Belgium: 97, 238, 242

Gillhausen, Gisbert: 100, 108, 115
Glasgow, Scotland: 120–21
Graf Zeppelin (airship): 270, 279
Gramme, Zénobe Théophile: 27, 154
Great Britain: 152, 175, 207, 281
Great Exhibition of 1851 (Great Britain): 15
Greer, Paul: 227
Grillon Frères breweries: 57, 62
Gripsholm (motorship): 171
Grosser, Karl: 136
Grusonwerk: 181–82
Guiberson Diesel Engine Company: 283
Güldner, Hugo: 183, 252

Hall, Charles Martin: 153
Hamburg-American line: 170–71, 202
Harland and Wolff: 170
Hartenstein (chemical engineer): 107
Hartmann, Wilhelm: 107, 126
Hartwig engine: 182
Harwich, England: 35–36, 244
Hautefeuille, Jean de: 33
Heilman (real estate firm): 235
Heine Safety Boiler Company: 178
Heldey, William: 30
Helmholtz, Hermann von: 153
Henschel Company: 271
Hermann Krabb (motorship): 270
Hermod (wooden steamer): 169
Hertz, Heinrich Rudolf: 32
Hesselman, K. J.: 168
Hindenburg (airship): 156, 270, 279–80
Hirsch, Baron Moritz von: 55
Hispano-Suiza works: 280

Hofmann, August Wilhelm von: 153

Holland, John: 156, 252

Howaldt Brothers (Gebrüder-Howaldt-Werft): 126–27, 202

Hubert, Captain H. (ferry master), holds his craft for search: 245

Hugo, Victor: 61, 64

Hvalon (gunboat): 213

Imperial General Staff: 222–23

Indianapolis Memorial Day Race: 258

Ingersoll-Rand Company: 259

International Harvester Company: 257–58

International Power Company of New Jersey: 253

Internationale Baufach-Ausstellung: 237

Ipswich, England: 229, 238, 240, 248

Isar River (Munich): 131, 157

Italy: 197, 207, 232, 275–76, 283

Jalbert-Loire (designer): 280

Jenatzy, Camille: 189

Joule, James: 153

Jung Company: 271

Junkers, Hugo: 278

Junkers Flugzeug und Motorenbau Gesellschaft: 278–79

Jutlandia (cargo vessel): 169

Kansas Pacific Railroad: 178

Kars (gunboat): 213

Kassel, Hesse: 122, 139–40

Kelvin, Lord: 121

Kempten, Bavaria: 127, 180

Kiel, Schleswig-Holstein: 191

Kiel-Gaarden: 183

Kiev, Russia: 214

Klöckner-Humboldt-Deutz: 269

Klüpfel, Ludwig: 100, 108

Koch, Robert: 152

Köhler, Otto: 77, 115; meets Diesel, 116

Kollektiv-Ausstellung von Diesel-Motoren: 132

Königliche Maschinen-bauschule: 115

Kosmos G.m.b.H.: 141

Krauss-Muffei Company: 271

Krumper, Josef: 77; belittles Diesel engines, 119, 120

Krupp Berlin-Tegel works: 183

Krupp, Fried., Werke: see Fried. Krupp Werke

Krupp, Friedrich: 28; Diesel contact, 77

Krupp Germania yards: see Germaniawerft

Krupp Südwerke: 268

La France (liner): 213

Lake, Simon: 156

Lake Starnberg (Bavaria): 42, 241

Lamm, O. (industrialist): 168

Lang, Franz: 278, 279

Langen, Eugen: 33, 76, 78, 184

Latouwski, Fritz: 219

Laukman, Alfred, goes to England with Diesel: 241, 244–46

Lauster, Immanuel: hired, 108; tested by Diesel, 109; represents factory, 136; directs engine work at M.A.N., 150

Legnano, Italy: 168, 207

Leipzig, Saxony: 77, 237

Le Tourneau Company: 259

Linde, Carl von: 48, 52, 55, 57, 58, 70, 78, 153

Linder, Hans: 93; gives silent salute to Diesel, 96

Lohner, Ludwig: 186

London, England: 13, 14, 15, 25, 26, 120, 167, 170

L'Orange, Prosper: 185, 188, 270; patents "pre-chamber" system, 188, 278

Lorraine Company: 281

Louisiana Purchase Exposition (St. Louis, 1904): 192

Ludwig, prince of Bavaria (later King Ludwig III): 133–34, 237, 241

Ludwig II, king of Bavaria: 237, 241

Ludwigsburg, Württemberg: 9, 10

McCarthy, J. L.: 226

McCarthy engine: 254

McCarthy, Norman: 251

McCormick, Cyrus: 154

McIntosh and Seymour Corporation: 256

McKeen, William: 217

McKeen rail car: 217, 219–21, 271

Mack Truck Company: 259

Magirus of Ulm: 268, 269

M.A.K. Company: 271

M.A.N.: *see Maschinenfabrik* Augsburg-Nürnberg

Manchester, England: 260, 275

Marx, Georg: 125

Marx, Karl: 12, 75, 92

Maschinenbaugesellschaft Nürnberg: 125, 178, 179, 252; exhibits engine, 132–34; merger of with *Maschinenfabrik* Augsburg, 150

Maschinenfabrik Augsburg: 57, 125, 136, 178; Diesel contact, 77; decides to construct Diesel engine, 78, 82; signs sales contract, 85; comments on engine, 90, 96; stock soars, 96; directors

meeting, 100–101; conference, 108; signs contract with Deutz, 116–17; franchise, 120; delivers first ship's engine, 127; exhibits engine, 132–34; amount spent by, 144; merger with *Maschinenbaugesellschaft* Nürnberg, 150

Maschinenfabrik Augsburg-Nürnberg (M.A.N.): 167, 271; merger, 150; engine sold by to Kiev, 180; litigation, 196; marine engines, 202; licenses Buda, 258; engine production of, 263–66

Maumee (naval vessel): 254

Maxim, Sir Hiram: 128

Maybach Company: 271

Maybach, Wilhelm: 83

Meier, Colonel Edward D.: goes to Augsburg, 125; makes report, 125–26, 178–79; education of, 177; army life, 178; lectures, 179; arranges Diesel's visit, 192, 194, 203, 211, 230; at party with Diesels, 238; designs submarine engine, 252

Meier, Hugh: 127, 179

Meistersinger von Nürnberg: 111

Meran, Tyrol: 146–47

Merceron, M. (engineer): 122

Metallwarenfabrik Berndorf: 102

Meyer, Paul: 136

Miller, Oskar von: 76

Mirrlees, Bickerton and Day: 274

Mirrlees, Watson and Yaryan Company: 120–21, 175, 274–75; Diesel's impatience with, 129; first Diesel electric installations in ships, for power plants, and battle tanks, 207

Mitterhoffer (inventor): 154

Monte Penedo (motor ship): 202
Morgan Construction Company: 252
Morse, Samuel F. B.: 32, 153
Motorenfabrik Deutz: *see Gasmotorenfabrik* Deutz
Motz, Lucie von: 74, 83
Münchener Allgemeine Zeitung: 96
Munich, Bavaria: 11, 46, 47, 48, 64, 113, 134, 180, 197, 239, 243
Munich industrial fair (1898): 182, 206
Münster, Graf, rejects gas as military weapon: 65

Nadrowski, Johannes (engineer): 90; makes engine drawings, 143
Napier, D., and Son: 281
Napoleon III (Louis Napoleon): 8, 15, 21; education of, 20; capitulation of, 24
National Aeronautics and Space Agency: 250
National Arms Factory: 283
Nederlandsche Fabriek van Werktuigen en Spoorweg-Materiel (Werkspoor): 167, 273–74; builds reversible engines, 207
Newcomen, Thomas: 29
New London Ship and Engine Company: 254
News-Bee (Omaha): 227
New York City, New York: 192, 208, 228, 234, 258, 260
New York American: 206
New York Auto Show (1930): 258
New York Morning World: 191
New York World: 206
Nobel, Alfred Bernhard: 153
Nobel, Emanuel: 127, 151; ob-
tains rights for Russia, 173; makes new agreement, 200
Nobel, Ludwig: 173
Nobel engine factory: 173
Noé, Ludwig: 123, 129, 136, 149, 266
Nordberg Manufacturing Company: 255, 257
North Pole (ice breaker): 213
Nürnberg, Bavaria: 13, 14

Old Mechanics Hall (St. Louis): 209
Oranje (liner): 272
Otto, Nikolaus August: 33, 76, 83, 184
Otto, N. A., & Cie.: 184
Otto Company (France): 139
Otto Engine Company (Philadelphia): 254
Otto gas engine: 76, 184

Page, Charles: 154
Panama Canal: 228, 249
Panama Fair (1914): 228
Papin, Denis: 29, 30, 198
Parerga und Paralipomena: 242
Paris, France: 5, 6, 7, 8, 15, 57, 97, 197, 236, 280; anti-German sentiment in, 22, 64, 67; world's fairs, 69, 70, 139, 151, 165, 184, 210
Paris Aero Show (1921): 283
Paris to Vienna race: 189
Parseval, Major: 132
Parsons, Sir Charles: 202, 238, 248
Paucksch, H. (manufacturer): 269–70
Pawlikowski, Rudolf: 136; forms own company, 141–42
Pennsylvania Railroad: 225, 226, 237

Penrose, H. J., holder of world's altitude record: 281
Perry (naval ship): 255
Petit Pierre (canal boat): 174
Pirrie, Lord: 170
Platt, Mr. (engineer): 120
Plunger (submarine): 252
Polytechnic High School, Munich: 46, 47, 48, 51, 86
Post-Dispatch (St. Louis): 226, 227, 229–30
Power and Mining Machine Company: 253
Pretoria (liner): 191
Prinzregent Luitpold (warship): 202
Prussian State Railways: 208, 214, 271
Puchta, Edward: 127, 179

Rathgeber Wagon Factory: 57
Reichenbach, C., *Maschinenfabrik:* 40
Reichenbach, Carl Ludwig: 40
Reichenbach, Fritz: 105, 136
Reichsbank: 237
Reisinger, Hugo: 179
Remscheid, Rhineland: 60, 241
Richard-Brasier racing car: 189
Riedlinger, L. A., *Maschinen und Bronzewarenfabrik:* 269
Rieppel, Anton von: 125
Riess, Sabine: *see* Diesel, Sabine Riess
Ringling Brothers Circus: 222
Rochas, Alphonse Beau de: 33, 50
Rocket (train): 260
Rotterdam, Netherlands: 36, 244
Rousseau, Jean Jacques: 21, 61
Royal Automobile Club of London: 240

Royal Shell Petroleum Company: 207
Ruppel, the Rev. and Mrs. E. L.: 229
Russia: 173, 225, 264, 265, 266, 267
Russian Diesel Engine Company: 128

Sachsen (dirigible): 237
St. Louis, Missouri: 124, 177, 192, 202, 209, 223, 226, 228, 271
St. Paul, Arkansas: 221
St. Petersburg, Russia: 127, 173, 200–201
San Antonio (clipper ship): 167
Sander, Ludwig: 39
Sauvage, Edouard: 122
Saurer, Adolph, Company: 272
Saurer, Hippolyt: 272–73
Savery, Thomas: 29
Saxon State Railways: 208, 271
Schelhorn, Johann Georg: 10
Schloss Labers (Meran): 147–49
Schmidt, Freiherr Arnold von (son-in-law): 189, 243; marriage of, 199; birth of first daughter, 235; visit of Diesel with, 241, 242
Schmidt, Hedy von: *see* Diesel, Hedy
Schmitz, Albert: 100, 108
Schmucker, Friedrich: 93; supervises engine work, 109; meets Meier, 125; installs first engine, 127
Schönbein, Christian: 153
Schopenhauer, Arthur: 242
Schröter, Moritz: 78, 117, 123
Schtorm (gunboat): 283
Schumm, Hermann: 115
Schweizerische Kohlstaubfeurungs A.G.: 141

Schwer, Kathrine von (maternal grandmother), forebears and death of: 14

Scott and Hodgson, builds first two-cycle Diesel engine: 275

Sears, Roebuck and Company: 258

Second Power and Works Machine Exhibition (Munich 1898): 130–34, 185

Seebeck, Johann Thomas: 32

Seib Brothers Woodworking Factory: 252

Seisel, Barbara (ancestor): 9

Selandia (cargo vessel): 169–70, 276–77

Shell Pipe Line Company: 255

Shenandoah (dirigible): 282

Shipbuilding Technical Union *(Schiffsbautechnische Gesellschaft)*: 236

Shole, Christopher: 154

Siemens, Werner: 28, 73

Smith, Leon: 216, 217

Société des Forges et Chantiers de la Méditerranée: 97, 122

Société des Ingénieurs Civils (France): 67

Société Française des Moteurs Diesel (Anciens Établissements Sautter-Harlé): 122, 174–75

Söhnlein, Julius: 115

Soleure, Switzerland: 272

Solvay, Ernest: 153

South America: 270, 274

South Kensington Museum (London): 26, 175

South Pacific Company: 252

South Pole: 168, 202

Southern Railways: 260

Soviet Russia: *see* United Soviet Socialist Republics

Sperry, Elmer A.: 256

Speyer, Rhein-Pfalz: 10

Ssarmat (tanker): 174

Standard (yacht): 169

Standard Fuel Engine Company: 256

Stanley, Henry: 154

Star (St. Louis): 229

Starnberger See (Bavaria): *see* Lake Starnberg

Stein, Carl: 115

Steinheil, Carl: 32

Steinmetz, Charles Proteus: 32, 154

Stephenson, George: 30, 260

Strategic Deterrence Force: 250

Strobel, Elise: *see* Diesel, Elise Strobel

Strobel Georg Friedrich (maternal grandfather): business of, 13; death of, 14

Stuart, Herbert: 251

Studies in Pessimism: 242

Stuttgart, Württemberg: 9, 99

Sulzer Brothers: *see Gebrüder Sulzer Maschinenfabrik*

Sulzer-Imhoof, Frau, reaction to Diesel of: 237

Sulzer-Imhoof, Jakob: 121–22, 176

Sulzer-Schmidt, Herr: 121

Sunbeam Motor Car Company: 282

Swabians: 9, 237

Swedish American Line: 171

Swedish State Railways: 168

Swiderski machine works: 77

Swiss Coal Dust Firing Corporation: *see Schweizerische Kohlstaubfeurungs* A.G.

Symington, William: 30

Tannhäuser: 159, 237

Technische Hochschule
  (Munich): 239
Tesla, Nikola: 154
Thaheld, F. A. (engineer): 283
Thery, Jacques: 189
Thirty Years' War: 9
Thomson, Sir Joseph: 153
Titanic (liner): 204, 230, 231
Tosi, Franco: 168, 254, 275–76;
  builds most powerful single-
  unit engine, 207–208
Trevithick, Richard: 30
Turin, Italy: 202, 283
Tüssel, Hanns (ancestor): 9
Tüssel, Hanns Jörg (ancestor):
  9, 10

Ulm, Württemberg: 10, 268
Union *Aktiengesellschaft,*
  Kempten: 127, 180
Union of German Engineers: 122
Union Pacific Railroad: 178, 260
United Soviet Socialist Repub-
  lics, Diesel engines in: 262–63,
  265
United States Naval Academy,
  Diesel lectures at: 208, 228
United States Navy: 252, 259,
  262
United States of America: 137,
  151, 177, 207, 274; Diesel en-
  gine in, 250–62; aircraft engine
  in, 282–83
United States Shipping Board:
  256

"Van Vou" automobile: 188
Vancouver, Canada: 261, 271
Vickers Sons and Maxim: 126
Vienna, Austria: 11, 148, 189
Vincennes, France: 22, 62
Vlissingen, Netherlands: 244;
  port records of, 246
Vogel, Lucian: Diesel contact
with, 77; assistant to Diesel, 86,
  89; represents company, 108;
  disagrees with Diesel, 119;
  meets Meier, 125
Vogt, Hans: 120
Voith Company: 271

Waldorf-Astoria Hotel (New
  York): 192, 234
Walter Company: 283
Wandal (steamer): 174
Watt, James: 28, 29, 121
Weiss, Carl W.: 251
*Werkspoor: see Nederlandsche
  Fabriek van Werktuigen en
  Spoorweg-Materiel*
Westfalen (vessel): 280
Westinghouse coach: 217
White Star Line: 204
Wilhelm II, kaiser of Germany:
  92, 171, 189, 222
Wilson, Alan: 217, 219–22
Wilson, Alf: 218
Winslow, Professor: 126
Winterthur, Switzerland: 52,
  121, 237, 272
Winton, Alexander: 255
Wheatstone, Sir Charles: 32
Woolson, Captain Lionel: 282
Woomera, Australia: 250
World Congress of Mechanical
  Engineers (London 1911): 202
World's Fairs: Brussels (1910),
  187, 207; Liège (1905), 167;
  Paris (1867), 184; Paris (1889),
  69; Paris (1900), 210; San
  Francisco (1915), 228, 249
World War I: 207, 256, 262, 263,
  267, 271, 273
World War II: 257, 268, 274, 279
Worthington Pump and Ma-
  chinery Corporation: 253, 254
Württemberg: 9, 10

Würzburg, Bavaria: 36

Z (submarine): 174
Zephyr (train): 255; first run, 260
Zeppelin airship: 229

Zeppelin, Graf Ferdinand von:
157, 163, 166; launches dirigible, 155; encourages Diesel to
build engines, 155
Zeuner, Gustav: 78